BIOTECHNOLOGY AND COMMUNICATION

The Meta-Technologies of Information

LEA'S COMMUNICATION SERIES
Jennings Bryant and Dolf Zillmann, General Editors

For a complete list of titles in LEA's Communication Series please contact Lawrence Erlbaum Associates, Publishers at www.erlbaum.com

BIOTECHNOLOGY AND COMMUNICATION

The Meta-Technologies of Information

Edited by

Sandra Braman
University of Wisconsin–Milwaukee

Copyright © 2004 by Lawrence Erlbaum Associates, Inc.
All rights reserved. No part of this book may be reproduced in
any form, by photostat, microform, retrieval system, or any other
means, without the prior written permission of the publisher.

Lawrence Erlbaum Associates, Inc., Publishers
10 Industrial Avenue
Mahwah, New Jersey 07430

Cover photograph by Graham Murdock
Cover design by Kathryn Houghtaling Lacey

Library of Congress Cataloging-in-Publication Data

Biotechnology and communication : the meta-technologies of information / edited by
Sandra Braman.
    p. cm. — (LEA's communication series)
    Includes bibliographical references and index.
    ISBN 0-8058-4304-3 (alk. paper)
    1. Biotechnology—Social aspects. 2. Communication. 3. Information technology.
  4. Information theory. 5. Bioinformatics. I. Braman, Sandra. II. Series.

TP248.23.B56 2004
303.48′3—dc21
                                    2003059933
                                             CIP

Books published by Lawrence Erlbaum Associates are printed on acid-free paper,
and their bindings are chosen for strength and durability.

Printed in the United States of America

for Anne Wells Branscomb
(1928–1997)

Contents

Introduction

Sandra Braman
University of Wisconsin–Milwaukee

Biotechnology, like digital technology, forces us to reconsider fundamental questions about the nature of life, human agency, and relations between the biological and social worlds. Many strands of communication research have examined issues raised by biology, but biotechnology pushes the research agenda of each forward into new terrain:

• The biology of human communication is a long-standing research area that now, as Wildman points out here (chap. 3), must include attention to the question of whether interventions wrought by biotechnology will affect the biological explanations for and constraints on human communication.

• Technological innovation and growth of knowledge about the gene have stimulated use of a shared vocabulary in discourses about biological and human information, although as Ritchie (chap. 2) makes clear, there are limits to the validity of the concept for each type of communicative process as well as problems raised by metaphoric transfer. At the same time, as demonstrated in two chapters—on conditional expectations (Wildman, chap. 3) and on facticity (Braman, chap. 4)—there is at least heuristic utility in exploring the implications of this shared discourse in some detail.

• The study of science communication has traditionally looked at factors affecting reportage about science and the role of such reportage in risk perception. The complexities of biotechnology suggest the need for multicausal analyses of reportage such as that offered here by Priest and Ten Eyck (chap. 7). Further, as explored in this collection by Murdock (chap. 9) and Best and Kellner (chap. 8), coverage of biotechnology is bringing media into new roles as active players in the debate over postnormal science and the democratization of decision making about uses of scientific knowledge.

• The convergence of computing and communication technologies has been the subject of extensive investigation. Yet as the opening chapter of the book points out, the growing convergence between information technologies and the organic world also requires the attention of scholars of information, communication, and culture.

• It has long been understood that communication is central to the sociology of knowledge. However, the disruptive character of contemporary innovations in biotechnology—as of digital information technology—is restructuring the institutions and practices of knowledge production and certification and the very nature of intellectual property rights, explored here by Lievrouw (chap. 6) and May (chap. 5).

This collection, limited as it is to a single volume, is far from exhaustive. A complete research agenda for biotechnology's impact on and implications for information, communication, and culture would also include the changing nature of individual and social identity, changes in organizational form and financial instruments, reconsideration of human communication processes as a result of what has been learned about cellular and biochemical communications, and legal and cultural implications of the merging of the digital and organic worlds. A brief review of the major themes in the history of biotechnology should help contextualize and focus the work presented here, and that which is to come.

Biotechnology is not a single technology, but a suite of techniques for processing genetic information derived from a number of disciplines, including biochemistry, molecular genetics, microbiology, and zymotechnology (fermentation). In its current form, biotechnology refers to processing technologies that apply microorganisms, cell cultures, or parts of either for human (industrial) purposes. It includes the design and use of microorganisms for direct use in food or other purposes (what economists refer to as a primary, or final, good) and genetic manipulation of microorganisms to improve their efficiency in converting materials that serve as inputs into other processes (a secondary good). Using biotechnology, genetic information can be processed either in a laboratory or in an organism (Goodman, 1987; OTA, 1982).

More simply, and covering a longer span of human history, the term biotechnology refers to any application of biology for human purposes (Goodman et al., 1987; Reiss & Straughan, 1996). Understood in this way, the practices of biotechnology are ancient; plant and animal breeding, and the use of yeasts for both fermentation and for leaching minerals out of rock, go back about 9,000 years. From its origins in such cultural practices, biotechnology slowly became codified as an explicit and shareable body of knowledge through a number of conceptual, methodological, and theoretical breakthroughs in systematic thought about the nature of life and its processes. These developments in turn effected and were affected by a growth in the capacity to store, process, and own genetic information.

The appearance of biology as a subject of study, the transformations of biology as a result of biotechnological developments, and the convergence of the biological with the mechanical all reflect and stimulate shifts in the understanding of the nature of life. Although the cultural practices of breeding and the use of microorganisms for human purposes are premodern, as a science biology is very much the product of modernity and "new" biotechnology and its products are the stuff of postmodernity.

The notion of life as an abstract concept, introduced by Cuvier at the turn of the 19th century, was a precondition of the possibility of biology (Foucault, 1978). Thus, it was only around 1800 A.D. that interest in biology as the study of the internal processes of life—as opposed to natural history as the study of the external forms of life—began to appear. Doing so raised the question of how species change, leading Lamarck to the realization that complexity and diversity could come from simplicity, and Darwin to turn from the tradition of scientific observation to experimentation as he tried to deliberately accelerate evolutionary processes in test tubes (Caron, 1988; "Evolution in a Test Tube," 1993; Harwood, 1993).

Knowledge expanded in several directions. First, although individual organisms were no longer seen as exclusively created through divine impulse, the lives of animals and plants were soon reconnected with the natural world through the concept of the biosphere, inspired in the 1870s by the study of relationships between biological and geological processes (Elichirigoity, 1999). Second, the individual organism as a whole dissolved into its parts and processes (Bud, 1993). Third, chemistry and then biology began to be seen as a way to link the interpretation of living processes with their technological—and commercial—exploitation (Guattari, 1992).

The consequences of these intellectual moves have been dramatic. For several decades, it has been possible to grow cells and tissues in the laboratory—outside of any living organism. The first transgenic species was patented in the late 1980s, and by now entirely new life forms are being created (e.g., Genentech's bug capable of making a protein foreign to itself; "Peering into 2010," 1994). The medical and artistic incorporation of technologies into humans, on the one hand, and the appearance of cognitive abilities, what appears to be creativity, and seeming self-consciousness and self-organization in electronic forms of artificial life, on the other, further challenge our understanding of just what life is and what it is not. Meanwhile an ever-growing proportion of the communications flowing through the global information infrastructure is the product not of humans, but of machines. The question of what distinguishes human communication from other types of information flows is no longer obvious.

Those involved in biotechnology have pioneered in the development of research methods more than once. Systematic experimentation was first undertaken by those fascinated by the microbe in the 1880s (Vernon, 1990). Completely aseptic laboratory conditions were the innovation of Chaim Weizmann (later the first president of Israel), whose biotechnological process for synthesizing acetone and butanol were critical both to the military during World War I and to those who needed convincing that the industry had commercial potential (Bud, 1993). The shift from designing experiments one by one to launching myriad processes to see which would be most successful was first undertaken in an analysis of genomes. Moving away from ancient breeding practices toward more aggressive hybridization based on specific characteristics as opposed to the health and/or desirability of complete plants or animals was a fundamental conceptual transformation that ultimately stimulated a number of other changes in research methods as well as in theory. Biotechnology was also very early to pick up on the importance of chance as a fundamental natural process shortly after it was discovered as a principle by those studying radioactivity. Of course biotechnology as a science has gone through several phases. Up until the 1970s, the field relied primarily on the use of natural organisms—"classical" biotechnology—but since then has turned toward manipulations of natural material and the ways in which it can be processed (Krimsky, 1991).

These developments were accompanied by the establishment of collections of materials on which to exercise the new research methods. During the late 19th and early 20th centuries, the first "archival" collections of genetic information were established. Most were specific to certain types of material such as fungi (The Netherlands, 1906) or microorganisms (Prague, 1884). During the same period, however, general collections were also established; Russia set up the Vavilov Institute, the oldest and still one of the most important seed banks in the world; and the U.S. Department of Agriculture set up a Plant Introduction office, institutionalizing the long-standing practice of aggressively collecting plant genetic material from around the world. Techniques for the long-term storage of genetic information through deep freezing came into use just around the time the problem of genetic erosion—the accelerating loss of genetic diversity—became a concern in the late 1950s and 1960s. Not surprisingly, an interest in establishing property rights accompanied the development of modern techniques of biotechnology and collections of processed material. Similarly, the commercial lure of biotechnology and the fact of its interdisciplinary nature nurtured efforts to receive recognition as a stand-alone discipline, on the one hand, and experimentation with organizational form, on the other.

The chapters of this book approach this complex history and the issues it raises from a number of directions. The opening chapter examines the shared features and spaces of biotechnology and digital information technologies as meta-technologies, qualitatively distinct from both the tools first used in the premodern era and the industrial technologies that characterized modernity. The next three chapters explore what is useful and what is not in treating the types of information processed by the two meta-technologies through a shared conceptual lens, each from a different perspective: Ritchie (chap. 2) takes a philosophical approach to the implications of the relationship between the tangible and intangible as suggested by references to the gene as information, Wildman (chap. 3) uses concepts from economics to look at the effects of conditional expectation in both genetically driven and human communication, and the chapter (chap. 4) on facticity is an exercise in narrative analysis. The next two chapters look at issues raised by the ownership of genetic and digital information, again from quite different perspectives: May (chap. 5) approaches the question as a legal problem, and Lievrouw (chap. 6) does so as a trend in the sociology of knowledge. The final three chapters are concerned with relationships between information and power, again from diverse positions: Priest and Ten Eyck (chap. 7) try to understand shifts in public opinion regarding genetically modified foods, Best and Kellner (chap. 8) look at the implications of debates over biotechnology for the emergence of postnormal science, and Murdock (chap. 9) analyzes the role of images in the struggle over genetically modified (GM) foods as they interact with cultural trends in response to digital information technologies to reach some conclusions regarding the relationships between postnormal science and the exercise of power. We hope this is just the beginning of the conversation.

Acknowledgments

Sandra Braman
University of Wisconsin–Milwaukee

The trajectory of this book marks the developing edge of a field. I first issued a call for participants in a panel on the relationships between communication and biotechnology in 1989, for an annual conference of what was then called the International Association of Mass Communication Research (IAMCR, now the International Association for Media and Communication Research). There were absolutely no responses to that call, leaving me feeling as if I were not on the cutting edge, but rather *over* the edge of a cliff even the face of which seemed to be invisible to others. It took about a decade before papers in this area began to appear within the field of communication, first and most often dealing with public opinion regarding biotechnology. Having jumped off the cliff in 1989, I continued to work on this topic throughout the 1990s and was able to finally present it publicly first in 1999 at both IAMCR and International Communication Association (ICA) conferences. Now, of course, there is a deluge of work, and it is hoped that the quite diverse chapters of this book—including analysis of public opinion, but also going far beyond that topic—mark points on a compass for a research agenda of multi- and interdisciplinary use.

The undergraduate library at the University of Illinois at Urbana–Champaign was built underground so as not to disturb what local lore describes as the first agricultural test plot in the United States. It was the agronomists of that university who first alerted me to the seed as biological information. Jennings Bryant must be thanked for his openness to the subject matter in his editorial role at Lawrence Erlbaum Associates, and Linda Bathgate of the press for her immediate and continued support for this project. The authors of this book undertook explorations that were in many cases diversions from or expansions on preexisting research agendas, providing the field with great gifts. Input from Christopher May and Steve Wildman provided particularly valuable spurs (and correctives) to my own thinking as the collection came together. Research assistance came from Mina Lee and Carol Ringo. The most enduring support for the fundamental premises of this endeavor across the years came from Peter Wissoker, whose subtle and active intellect is deeply appreciated and to whom a debt is clearly owed. Without Guy W. Milford's extraordinary support for all aspects of this work, from the most abstract to the most mundane, the book would not have been possible.

The book is dedicated to the late Anne Wells Branscomb, who served as mentor and role model from our first meeting in 1985. Anne was working on issues raised by biotechnology as an information technology at the time of her death in 1997—as usual, leading the way with keen intellectual insight and inestimable grace.

THE TECHNOLOGIES
OF BIOLOGY
AND COMMUNICATION

1

The Meta-Technologies of Information

Sandra Braman
University of Wisconsin–Milwaukee

Against the long history of the use of tools and technologies, contemporary biotechnology and digital information technology together fall within a third category—meta-technologies. Their shared meta-technological characteristics make it worth examining them side by side because they often share economic, social, cultural, and legal spaces. These shared characteristics will become even more important as digital information technologies and biological organisms merge, for the convergence of technologies that brought computing and communication technologies together is just one in a series of types of convergence of communication with other materials and social processes. There have been four:

1. The convergence of symbolic communication with materials when language was first expressed in writing.
2. The convergence of symbolic technologies with those of energy in the mid-19th century, launching the information society.
3. The convergence between computing and communication technologies made possible by digitization in the mid-20th century.
4. The convergence between digital technologies and the organic world, including the human body.

This chapter looks at the nature of meta-technologies, explores the shared spaces of digital information technology and biotechnology, and suggests what the implications of the shared features and spaces of meta-technologies might be as digital technologies and organisms increasingly converge. Of course biotechnology and digital information technology, and the types of information they handle, are also very different. The goal here is thus to be suggestive and, hopefully, provocative in ways that should stimulate further thought and research of value to both fields.

META-TECHNOLOGIES

All along the invention of new kinds of tools and technologies has had such an impact on the nature of society that we distinguish among the premodern, modern, and postmodern periods according to that which has dominated in each. The specific dimensions along which informational meta-technologies differ from industrial technologies go far toward explaining why the current period is also described as an information society. Indeed, although the point should not be overdrawn, there are some interesting parallels in developments in both types of meta-technologies at different stages of the history of the information society. The characteristics of meta-technologies also explain why the convergence between technologies and organisms is accelerating.

Meta-Technologies

The word *technology* has its roots in the Greek *techne*—"making"—referring to what both art and engineering have in common. Three different ways of "making" have developed: the ancient tools of premodernity, the industrial technologies of modernity, and the informational meta-technologies of postmodernity.

Tools. Tools can be made and used by individuals working alone. They process matter or energy in single steps. The use of tools characterized the premodern era. Although it is easy to think of examples of ancient tools for other things people do, like planting seeds or starting a fire, because communication is an inherently social act it may only be when marks are made for the purposes of individual memory can it be said there are communication tools. The use of yeast for brewing, believed to be 9,000 years old (Krimsky, 1991), would be an example of a biotechnology tool because it merely involves introducing a natural yeast into a mixture of organic materials.

Technologies. Technologies are social in their making and use, requiring a number of people to work together. They make it possible to link several processing steps together in the course of transforming matter or energy, but there is only one sequence in which those steps can be taken, only one or a few types of materials can be processed, and only one or a few types of outcomes can be produced. The shift from tools to technologies made industrialization possible, and the use of technologies thus characterizes the modern period. The printing press and the radio are examples of communication technologies. Using fermentation to synthesize materials in a laboratory is an example of a biotechnology technology.

Meta-Technologies. Meta-technologies involve many processing steps, and there is great flexibility in the number of steps and the sequence in which they are undertaken. They can process an ever-expanding range of types of inputs and can produce an essentially infinite range of outputs. They are social, but permit solo activity once one is operating within the socially produced network. Their use vastly expands the degrees of freedom with which humans can act in the social and material worlds, and characterizes the postmodern world. Meta-technologies are always informational, and the internet is a premiere example of a meta-technology used for communication purposes. With recombinant DNA, biotechnology entered the meta-technology realm.

The change in human capacity enabled by meta-technologies is both qualitative and quantitative in nature. It is accompanied by a loosening of historical constraints on decision making about production and other social processes. Some types of path dependency and structural constraints can now be side-stepped altogether. Because the range of possibilities is so much greater than before, what has been learned in the past about how to make decisions does not always suffice. The underlying premodern and modern assumption that an equilibrium can be achieved—that there is a right answer—is irrevocably gone.

Dimensions of Difference

Tools, technologies, and meta-technologies differ along four dimensions—the degree to which they are social, the complexity of the processes they enable, their autonomy, and their scale—with the movement from tool to technology to meta-technology marked by an increase on each of these dimensions. Other features are also worth noting.

Buckminster Fuller (1975) introduced the notion of the social nature of technologies when he discussed writing as the first technology. The social coordination required for the use of technologies and meta-technologies explains why it is so important to agree on both technical standards and protocols for their use. It also explains why their use has such an impact on society because each requires or enables the development of specific types of coordination and interaction.

Marshall McLuhan (1964; McLuhan & Fiore, 1968) drew attention to the second feature, complexity, when he noted that both tools and technologies change the field of possibilities and therefore of practice. French philosopher of technology Jacques Ellul (1964) offered a more detailed way to think about this when he defined *technique* as "a complex of standardized means for attaining a predetermined result" (p. 4). Complexity is a feature not only of entire processes enabled by specific technologies, but also of each of the steps of which such processes are comprised; the more complex, the greater the possibility of flexibility and creativity (Novak, 1997; Scazzieri, 1993), although an increase in complexity does not always mean a better technology. The only limits to the complexity of digital meta-technologies are those of mathematics and imagination; we do not yet know the limits of biotechnology.

Concern over the autonomy of technologies appeared first in the 11th century in the Golem stories that later inspired *Frankenstein*. These tales of a creature made out of clay to serve human needs always concluded with the Golem becoming destructive because people were unable to be sufficiently detailed and accurate in their instructions. The notion of machinic autonomy, defined as technological agency outside the limits of human control, thus appeared early in the transition from tools to technologies. Economic historian Chandler (1977) points out in his seminal work, *The Visible Hand*, that beginning with the automation of production lines in the 19th century, society began turning its decision making over to machines, thus granting machines a second type of autonomy. In the digital world of meta-technologies, a third type of machinic autonomy has appeared in intelligent agents that roam the networks finding information, making decisions, and conducting transactions of their own on behalf of humans. It may go even further: Non-human network intelligences are now making decisions on their own—and increasingly our (Braman, 2002b)—behalf. Dyson (1997) points out that machinic intelligence may now be operating autonomously in ways that humans cannot even perceive because the logics may be so different from what we know.

Of course globalization and many of the impacts of the information economy are the result of vast increases in the scope and scale of activity. The same thing can be said about recent developments in biotechnology. As discussed in more detail later, the numbers of combinations and manipulations of genetic information that can be analyzed and the speed with which such analyses can be iterated has so increased that the very nature of scientific practices has changed. So too there has been a change in scale in the quantities of products produced by biotechnology. Biowarfare, for example, has risen in concern because while before noxious substances were only available on a small scale, it is now possible to make such substances in any quantity desired, completely changing the nature of warfare (van Creveld, 1991).

Other features of meta-technologies also distinguish the modern from the postmodern era. Belief that technological development was always progress was a hallmark of modernity, whereas in the postmodern world growing concern about technological risk has stimulated the debate over postnormal science discussed by Murdock (chap. 9, this volume) and Best and Kellner (chap. 8, this volume). Technologies used to be viewed as stand-alone objects, while today it is understood that each is inextricably part of a system. We used to apply the term technology only to material objects, but now we use it to refer to ideas and ways of doing things as well. Despite the modern fancy that technology and culture have little to do with each other, it is clear that each deeply informs—indeed creates—the other.

The Information Society

Although meta-technologies did not come into widespread use until fairly recently, there are provocative parallels between the development of technologies for the biological and communicative realms at each stage of the history of the information society. Of course it is commonplace to note that information was important to society long before the concept of the information society appeared, and historians have begun the valuable work of demonstrating just how this was so (see e.g., Chandler & Cortada, 2000; Headrick, 2000). There are a number of moments one could identify as the beginning of the informatization of society; however, the point at which communication technologies became electrified is a useful marker because from that point on the pace of change accelerated and movement through high modernity and then postmodernity began (Braman, 1993).

The parallels actually began even earlier. Widespread diffusion of print and what is referred to as the "Columbian Explosion" (Crosby, 1972; Kloppenburg, 1988) or "Columbian Encounter" (Crosby, 1994)—the massive global flows of genetic information that were launched with Columbus' first visits to the Western hemisphere—were both late-15th-century phenomena. As Eisenstein (1979) and others show, effects of the printing press included an enormous stimulus to the development of knowledge because it was easier to transport information, compare information from one source to information from another, collect large bodies of information in one place, and of course reproduce it. Print also facilitated the standardization of weights and measures, spelling, and other matters important to the development of science. The bureaucratic forms it enabled encouraged a more finely articulated division of labor. Similarly, the Columbian Explosion transported enormous amounts of genetic information around the globe and made it possible to collect large amounts of germplasm in single sites and thus to compare and reproduce it. These capabilities were used by imperial governments to restructure the global division of agricultural labor and thus the global economy. In many colonies, complex and diverse ecologies and agricultural systems were demolished and replaced with monocultures devoted to producing single commodities to enhance the profits that might be gained through maximizing what economists call the "comparative advantage" of each—on behalf of the imperial center—via international trade.

Stage 1—Mid-19th Century. The communication technology that marked the first stage of the information society was the telegraph; it was invented in the 1830s, came into use in the 1840s, and was in global use by the 1860s. The telegraph differed from earlier communication technologies not only because it was electrical, but also because it "packetized" information into binary form (short or long) for transmission through the electrical circuits of the network. Communication via today's global information infrastructure of course still takes place in digital form (0s and 1s). In the meta-technological environment, it is the packetization of information that makes possible many processing capabilities; in transmission, messages are often broken into packets for delivery along separate routes, only to be recombined at the point of reception.

It was in the early to mid-19th century that organisms similarly became packetized, in the sense that the concept of the gene appeared and became refined. The result was a shift in perception of plants, animals, and people as holistic entities to seeing them in terms of their parts. In the meta-technological environment, the packetization of the organism remains important because it makes it possible to conceptualize and operationalize the recombination of genes in novel ways that are at the heart of the contemporary biotechnology industry.

Both types of information underwent classification during this period. There was another—trivial—connection between the two: Samuel Morse, inventor of the telegraph, traveled to Asia to collect soybean germplasm for the U.S. government.

Stage 2—Turn of the 20th Century. Both communication and the biological roots of what we now call biotechnology went through the throes of the effort to gain disciplinary recognition during the second stage of the information society. As part of that effort, both also became more systematic in their treatment of information—in biotechnology via experimentation and in human communication via explicit attention to the practices of those who work with information (librarians, journalists, accountants, etc.) and the professionalization of those practices.

Stage 3—The 1960s. Perhaps the most vivid parallel between the two appeared at the point of transition from industrial technology to informational meta-technology during the third stage of the information society, loosely ascribable to the 1950s and 1960s. By that point the convergence of computing and communication technologies that began to be possible during World War II had diffused broadly enough that its effects were widely experienced. The first legal problem raised by this convergence appeared in the mid-1950s (Pool, 1983), and self-consciousness regarding the impact of these new technologies on society and deliberate experimentation with their social effects shortly thereafter.

It was in the same period that the spiral helix of DNA was discovered, and by 1972 the processes of recombinant DNA became available. This marked the transition between "classical" and "new" biotechnology, the shift from working with germplasm in its natural form to an emphasis on processes of intervention (Van Wijk et al., 1993). Other differences between the two approaches echo features familiar to those who study digital technologies:

- Hybridity: Although there has been experimentation with breeding across species since ancient times, the extent of such experimentation has now increased, and the gap between species involved continues to widen, even extending across the plant–animal divide.

- Speed: The pace of change in traditional biotechnology was much slower; with new techniques, genetic information of one species can be permanently inserted into another within a few weeks rather than working on a scale of years. Millions and sometimes billions of copies of genetic information can now be made in just a few hours.

- Scale: Traditional biotechnology was only used on a relatively small number of species for limited purposes, whereas efforts today are much more ambitious, extending beyond food and drink to pollution control, sewage disposal, drug production, and fundamentally changing not only animals but humans.

- Precision: Traditional breeding methods transfer and recombine large numbers of genes in a largely random manner, making it difficult or impossible to predict with accuracy which or how many traits are transferred by these methods. With contemporary biotechnology techniques, however, it is possible to introduce individual pieces of DNA with great precision, yielding control over discrete genetic changes (Miller & Huttner, 1995).

Stage 4—The 1990s. Parallels between biotechnology and digital information technologies during the fourth stage of the information society are the subject of the rest of this chapter, but it should be noted here that the possibility of a convergence between machinic and biological technologies was first suggested by Mumford (1934) in his book, *Technics and Civilization*.

SHARED SPACES

Today the meta-technologies of biotechnology and digital information technology share a number of economic, cultural, social, and legal features and environments. The shared spaces of the two types of meta-technologies are evident in discourse, economics, culture, social processes, the law, and finally in the convergence of genetic and digital information.

Discourse

Discourse-based parallels between the information of human communication and genetic information appear in the course of description, in the concept of the media, and in rhetoric about each. At one point they were even linked as content: In the 19th century, agriculture journals gave seeds away with subscriptions (Kloppenburg, 1988). The limits to such parallels noted by Ritchie in this volume, however, should not be forgotten.

Description. By the mid-1980s, the language of biochemistry was filled with terms used in the analysis of human communication such as "recognition," "high fidelity," "messenger DNA," "signaling," and even "presenting" (Hoffmeyer, 1997). The genome is described as a complex parallel-processing computer or network, and some genetic information acts as switches—just as in the telecommunications network—to turn on other DNA. Although originally DNA was thought to be a collection of recipes for building proteins, it now looks more and more like a software program, embodying abstract symbol-manipulating machinery (DeLanda, 1991). The loss of species, therefore, is "an irreversible loss of information" (Bullard, 1988, p. 220). As Boyle (1996) put it,

> We have already reached the point where genetic information is thought of primarily *as* information. We look at the informational message—the sequence of As, Bs, Cs, and Ts—not the biological medium. The human genome project is simply a large-scale exercise in cryptography. Like archaeologists with the Rosetta Stone, we have broken the cipher, and can now deal with DNA as a *language to be spoken,* not *an object to be contemplated.* (p. 4; italics original)

Interestingly, the history of the treatment of germplasm as information has repeated some of the history of the treatment of the concept of information in human communication, such as the distinction between isolated bits of data and information that coheres into a narrative story (Oyama, 2000).

Mediation. Sunderland (2002) views biotechnology as a form of media because of its role in literally shifting and politicizing meanings. She argues that biotechnology involves four processes—alienation, translation, recontextualization, and absorption—that effectively influence and thus mediate our understandings and experiences of ourselves, other species, and the world in which we live.

Editorial and gate-keeping functions are among the key functions of media. Gene technology and media technologies such as those involved in screen editing and special effects both facilitate the fixing of information and representations thereof (Van Dijck, 1998). The morphing techniques made possible by digitization are considered genetic rather than surgical in that they are more like genetic cross-breeding than transplanting, blending the unfamiliar (Novak, 1997). Biotechnology can even now perform a "global search and delete" function in which genes of a particular group act like molecular scissors, cutting DNA molecules wherever a particular sequence of DNA "letters" appears ("Exterminate," 2003).

Some analytical and editorial techniques are now literally being used in common: Software designed to analyze microbial evolution is now being used to examine variants of texts such as the multiple variants of Chaucer's tales. Doing so is changing views of these texts because it was discovered that some lesser known variants may be closer to the original than those in standard modern editions. Repeated copying introduced ever-more errors, ultimately producing distinct versions akin to new species of life (Brainard, 1998).

Rhetoric: Utopia Versus Dystopia. Both genetic and digital information have been the subject of rhetoric at the utopian and dystopian extremes. For the information society, the first of these involved claims that digital communication would lead to democracy, erasure of socioeconomic class lines, a reduction in the time spent on work, and so on. The latter focused on reification of socioeconomic class differences, homogenization of content, loss of privacy, deskilling, loss of knowledge, a decline in organizational productivity, and even health and environmental problems.

For biotechnology the utopian version has appeared in claims since the early 20th century that it would address all nutritional deficits (Wilkinson, 1987), launch a new industrial revolution (Bud, 1993), solve all of the problems of the industrial world (Roobeek, 1990), and do so in the most "natural" and inevitable of ways. The dystopian response has noted that promises after 100 years remain unfulfilled, the use of biotechnology can result in decreased yields, and biotechnology causes environmental, health, social, and economic problems (Kloppenburg & Burrows, 1996).

Economics

The commodification of information is driving common economic trends in both the digital information and genetic information worlds—marginalization of producers, oligopolization, and financial innovation.

Commodification of Information. One of the key reasons the phrase
"the information economy" has come into use is that the very domain of the
economy has expanded through commodification of forms of information
never before commodified. As applied to digital information, this has meant
turning information that was both historically public (e.g., databases developed by the government to serve public purposes) and deeply private in
nature (e.g., attention) into products that can be bought and sold. As applied to genetic information, this has meant the progressive establishment
of property rights in forms of information historically considered resources
common to all humankind, to all within a particular society, or to an individual. Kloppenburg (1988) makes it profoundly clear in *First the Seed* that with
this shift capital finally penetrates into the most ancient of cultural habits
and those the most fundamental to survival. From the viewpoint of the law,
this was accomplished through a steady expansion of intellectual property
rights over various forms of genetic information, beginning with asexually
breeding plants in the 1930s, through sexually breeding plants, microorganisms, and so on, ultimately to include transgenic animals by the close of the
20th century. By today, as a result of these legal developments, medical researchers have the right to patents on materials or processes derived from
your own organs (Boyle, 1996), companies in the agriculture-food-chemical-
pharmaceutical industry can own patents to entire species of basic food
crops, and governments are attempting to stake out ownership of the genetic information not only of their own, but also in some cases of other peoples. In each case, extension of intellectual property rights to an additional
form of genetic information has triggered an explosion in commercial activity (Krimsky, 1991).

This commodification process has brought to light another feature
shared by digital and genetic information—their dual nature as both public
and private goods. Of course there are two meanings of public good. Politically, a public good has positive value for society as a whole and thus
should be accessible to all. Economically, a public good is nonexcludable
(potential users cannot be excluded from use) and nonrivalrous (one person's use of the good does not keep others from using it). Digital information is clearly a public good in the second sense and considered by many to
be a public good in the first sense, yet it is often treated as a private good
as a result of its embedding within or reliance on material goods that can
be owned through the legal creation of intellectual property rights. The
struggle over just where the limits should be between that which is private
and that which is public has resulted in one of the most hotly contested debates over approaches to regulation of the global information infrastructure and the content that flows through it.

Genetic information, too, is considered by many to be a public good in
both senses of the phrase, yet it is increasingly being treated as a private

good through the construction of property rights via the law. In the past, wild genetic resources have been available to all nonexclusively and non-rivalrously. As the intellectual property rights system develops to privatize more and more forms of genetic information, however, notions such as "farmers' rights," recognition of the value of landraces (genetic information adapted to particular environments over long periods of time) (van Wijk et al., 1993), and appreciation for the role of traditional forms of cultural knowledge as key to the use of genetic information are being developed to justify placing boundaries on the extent to which genetic information may be made private.

In both cases, alternatives to completely transforming public good information into private goods are being developed; these take advantage of contractual limits and conditions and political activism on behalf of the public interest. The notion of an information "commons"—a pool of information held in common by all the world's people and available to all—is being vigorously put forward by NGOs. (Because of the importance of Linux, the open source software, in educating people to the value of a commons, Srinivas [2002] uses the term "biolinuxes" to refer to the same concept as applied to genetic information.) The UN's Food and Agriculture Organization (FAO) first announced that plant genetic resources should be treated as a heritage of humankind and should be available without restriction in 1983, and this view was reinforced by the UN Convention on Biological Diversity. Principles must be turned into programmatic realities to have any impact, however. Experimentation with concrete legal and economic techniques to transform the concept into a logistical reality is underway. In policy discussions, debate over the commodification of information often takes the form of discourse over the relative merits of profit and the public interest as dominant decision-making values.

Marginalization of Producers. It is one of the ironies of the digital environment that precisely as the role of creativity has finally been recognized as significant not only for its cultural but also for its economic value, individual producers have become marginalized to the advantage of large institutional producers. In some cases, the processes of invention and innovation are so complex that they can only be undertaken within the context of large institutions with many resources, both human and otherwise, to devote to the problem. The result in such instances is that the work of individual producers is "work for hire" and thus the property of the hiring organization. In other cases, individual producers are forced to yield up intellectual property rights in their work in exchange for reproduction, marketing, and distribution, often netting very little—or even nothing—financially as a result of the exchange. Even the organizations of the information infrastructure are now claiming the right to work carried through the global network; "terms of service" and "acceptable use agreements" of ISPs increasingly assert a compulsory license in all material sent through their systems, including the right to use such material without permission, but with the author's name, for commercial purposes (Braman & Lynch, 2003).

Farmers and breeders have historically been the individual producers of genetic information, but their relative power is diminishing both because their numbers are dropping as a result of the automation of functions (Schuh, 1986, 1989) and because their relative economic role is small compared with that of institutions that process and distribute what they produce (Busch et al., 1991; Hadwiger, 1982). In 1776, 95% of the U.S. population was farmers, whereas by the mid-1990s, the percentage had dropped below 1%—so low that the U.S. Census Bureau announced that farmers as a group were no longer statistically significant and thus would no longer be tracked (McKibben, 1996). This drop in relative economic importance translates into a spiral of declining ability to negotiate for protection of producers' work through farmers' or breeders' rights.

Oligopolization. The trend toward oligopolization in the information industries involved in human communication has been the subject of scholarly analyses (e.g., Herman & McChesney, 1997) as well as lawsuits (e.g., the series of cases between the U.S. Department of Justice and Microsoft). Several factors have combined to bring about the same trend among firms that deal with genetic information. Beginning in the 1970s, rising grain prices, declining rates of profit in the chemical industry, extension of intellectual property rights to the products and processes of manipulations of the genetic information of sexually reproducing plants, the desire to "rationalize" agro-input marketing, and the importance of the export of agricultural products all encouraged multinational corporations (MNCs) and transnational corporations (TNCs) to buy up firms starting with feed companies and including every stage of the processing, marketing, and distribution chain (Mayer, 1986). The process accelerated when, in the 1990s, a number of important chemical and pharmaceutical patents expired at the same time that the ability to assert property rights in "new" genetic information products expanded; in combination, these factors changed the competitive nature of the market altogether.

In response, in 1994 alone, large U.S. pharmaceutical firms bought up 117 ventures with biotech firms—70% more than previous year ("Unseemly Couplings," 1995). This has not been just a U.S. phenomenon—between 1993 and 1995, around $70 billion worth of mergers and acquisitions took place within the European chemistry industry, which has a yearly turnover of just $200 billion ("Carving up Europe's," 1995). Around the world, the largest firms are consolidating in waves of mergers (Powell, 1996). The formerly distinct industries of agriculture, food, chemicals, and pharmaceuticals have come together into a single agriculture-food-chemical-pharmaceutical industry as a result of vertical integration in treatment of genetic information. The chairman of Del Monte, the largest processor of fruits and vegetables in the world, put it this way: "We literally begin with the seed and end at the grocer's shelf" (quoted in Mamiya, 1992, p. 49). As Boyle (1996) points out, monopoly property rights now being given to biotechnology and software companies rival anything given to railroad or banking trusts 100 years ago.

Financial Innovations. Dramatic IPOs (initial public offerings on the stock exchange) were a famous and striking feature of the dot.com boom in the information industries. The first highly publicized IPO, however, was in biotechnology, and it was in biotechnology that the first rash of such offerings—and the publicity that accompanies them—took place. For both industries, support from the investment community had a significant effect on the structure of the competitive field.

There were several phases of the investment boom in biotechnology. These began when Genentech was founded in 1976 with an IPO that received an enormous amount of media attention and set a record for a rise in price on the first day of availability. This was a new kind of IPO for Wall Street and seized its imagination. The business involved a highly exotic substance—even esoteric—leaving a lot of room for salesmanship. A few analysts developed such an aura around biotechnology that the stocks sold even though experts were undecided about the viability of the new processes and industry. Although no profit was expected for years, enthusiasm was so high that price setting was arbitrary, just as happened later with the dot.com boom. In the months that followed, dozens of similar companies also presented IPOs, generating a phenomenon so strong it took on the name of "biomania" and reaching a first climax in 1983. At that point, announcement of the oncogene, believed to be involved in the development of cancer, triggered a second, "heroic," rash of investment speculation, coaxing forth "exotic corporate life forms" (Teitelman, 1989, p. 93). By the end of the 1980s, however, there was a "loss of innocence" on the part of investors, who began to realize that—as was the case with the information industries—no profit was to be seen, and what had seemed so simple in fact was very complex. Still, massive amounts of money were being absorbed by nonprofitable enterprises in which investors regained confidence over and over again throughout the 1980s and 1990s ("Panic," 1994).

Public financing had significant impact on the biotechnology industry— and, interestingly enough, in doing so encouraged the trend toward postnormal science. In an environment in which decision making about medical funding had been controlled by the government via its funding mechanisms, investment financing decentralized decision making and opened it up to the market. The involvement of investment bankers, brokers, analysts, and the larger scientifically unsophisticated public in research decision making altered not only the rules of the game, but also the kinds of projects financed and pursued (Teitelman, 1989).

Other innovative financial mechanisms were stimulated by biotechnology. As farmers found their profits dropping and the safety net dropping away, many turned to derivatives and complicated stock options as a way to hedge price risk ("Old MacDonald," 1995). Revised tax laws made possible new types of financing packages (Krimsky, 1991). Even the London Stock Exchange was persuaded by biotechnology firms to drop its rule that there must be 3 profitable years before listing ("Biorhythms," 1993).

Culture

Areas in which there are resonances—and interactions—between the cultural manifestations and effects of biotechnology and digital information technology include their impact and expression in individual and national identity, the growing role of risk, and questions of cultural diversity.

Individual Identity. Digital information has made it possible for individuals to experiment with individual identity, strengthen traditional ethnic ties, and develop more expansive, multiple, and hybrid forms of individual identity. Genetic information is popularly considered to be the essence of identity, explaining individual difference, the moral order, and human fate—despite empirical facts to the contrary. It is believed to be incapable of deception and the locus of the true self, the secular equivalent of the soul. This "genetic essentialism" (Nelkin & Lindee, 1995, p. 2) reduces the individual to a molecular entity isolated from social, historical, and moral factors and turns the genome into a text. Paradoxically, however, the desire to treat the gene as the ultimate arbiter of human nature has increased at the same time that scientists have come to view the gene as both profoundly unknowable and changeable. As Gould (1997) notes, genetic essentialism also confuses correlation for causation.

Other challenges to human identity are raised by this attitude toward the gene: The long history of privileging the human must face a situation in which more is known about the genetic makeup of the worm than of any other animal (Kiernan, 1999; Wuethrich, 1993), and it is clear that the DNA of corn and salamanders is more than 30 times as complex as that of humans (Rabinow, 1996). The belief that the gene is immutable and determinant runs counter to contemporary scientific and popular views of the immune system as a chaotic, hyperflexible site ridden with contradictions and warfare (Appadurai, 1993). Artists have started using the interaction between genetic information and identity as a medium, as when a conceptual artist offered a Genetic Code Copyright (Nelkin & Lindee, 1995) or in genetic modifications as art (Dickey, 2001).

National Identity. Like the use of digital information technologies, biotechnology has been used to both undermine and strengthen national identity. Significant differences in the kinds of questions asked about genetic information make clear that attitudes toward biotechnology are inevitably culture-bound (Harwood, 1983). They are also political: Although it is claimed that ethnic identities are genetically based, in fact none of today's ethnic categories existed before the development of the capitalist world system (Quijano & Wallerstein, 1992). Under monarchies, complex intermarriages supported the exercise of power across different population groups within societies (Anderson, 1983), and the assertion of ethnic categories was critical to the history of colonialism (Desrosieres, 1998). Nation-states can contract for ownership of or control over domestic genetic information resources with those local entities that have historically had control, assert national ownership of and control over unprocessed domestic genetic information resources, claim control over biotechnologies and any products of biotechnological interventions invented by their citizens, and/or treat the biological information of its citizenry as a resource.

Control over the genetic information of indigenous animals and plants, long assumed, is now being actively asserted as a form of national power. The Biodiversity Convention of 1992 stressed that nation-states had control over their own genetic resources. Iceland has gone further, volunteering a complete analysis of the genetic makeup of its citizens. Various subnational cultural groups are reasserting the right to control the use of the landraces they have traditionally grown. Even shy of official geopolitical structures, genetically based relationships have historically been and remain extremely important economic structures (Thorbecke, 1992). Genetic information has also been sacred to many traditional cultures (Cleveland et al., 1994).

Although an insistence on government control over genetic information can serve a society by ensuring that it derives some benefit from the use of its resources, it can also result in difficulties. Japan, for example, only recently put aside a Staple Food Control Act that forbade the import of rice and protected traditional types of rice and farming practices in the name of national security even though doing so caused the price of Japanese rice to rise to 900% of the world market price. Under Pol Pot and the Khmer Rouge, only community (read, "modern") varieties of rice could be used, exacerbating widespread starvation because farmers were prevented from using the traditional deep water rice that grows on flooded land ("Not a Grain of Truth," 1992).

Cultural Homogenization. Cultural homogenization is one of the most common complaints about the effects of the globalization of human communication systems that has been so exacerbated with digital information technologies. Cultural homogeneity—known as monoculture when applied to agriculture—is also a consequence of the use of biotechnology; it is of enormous concern not only because of its effects on human culture, but also because of its environmental costs and the vulnerabilities it induces. The greatest damage that has been wrought through human manipulations of genetic information are not those from the latest rounds of recombinant DNA, but from the results of the Columbian Encounter, when massive global flows of genetic information were used to replace biodiverse ecologies with monocultures intended to serve economic rather than survival or cultural concerns. Although monocultures may increase profits during some periods, they can have devastating effects, as in the Irish potato famine. No society is impervious. A 1972 National Academy of Science study described crops as "impressively uniform genetically and impressively vulnerable" (Kloppenburg, 1988, p. 287). One of the first communications between the United States and Russia after the fall of the Soviet Union was a plea for the more genetically diverse cornstock in Russian hands, which was desperately needed to replant U.S. fields that had been devoted to monoculture and were devastated by disease (Strobel, 1993).

The destruction wrought by monoculture is also cultural and spiritual, as ancient relationships among social, agricultural, and religious practices are disrupted by commodification of the seed (Kloppenburg, 1988). Even in societies of the North, it leads to the loss of a particular type of lifestyle. As McKibben (1996) saw it, for several thousand years, one of the most important countervoices to a uniform material culture came from those involved with farming who were independent, planned for the future, involved with the natural environment, and lived where they worked.

Risk. The growing sense that the scientific and technical developments described as "progress" were in fact introducing additional elements of risk into society marked the beginning of the end of the modern period of society (Beck, 1992; Douglas, 1992; Douglas & Wildavsky, 1982; Rabinow, 1996). This notion first appeared in conjunction with digital information technologies in the late 1970s, when a report to the Swedish government on the computerization of society focused on the new types of vulnerabilities such technologies induced (Tengelin, 1981). Today the susceptibility of the information network to viruses and hacking, along with fear of information warfare make risk a central theme. Since the 1970s, risk has also dominated discourse about biotechnology. In both cases, both processes and products are of concern.

The possibility of biotechnology as a source of risk became public following the 1974 Asilomar conference at which a number of scientists made their own hesitations public. They did so just as the amount of government funding had begun to rise, and the fact that those scientists who stood to gain the most were the first to raise an alarm generated a lot of publicity. During the same period, the environmental movement was bringing attention to the decline in genetic diversity, another type of risk to which it was believed by many biotechnology would contribute (Mooney, 1988).

For biotechnology, drama—as discussed by Golding in his chapter in this volume—has contributed to the sense of risk. One of the first scares, in 1982, was generated by a request for deliberate release of a soil bacterium that seemed to come directly out of Kurt Vonnegut's (1963) novel *Cat's Cradle*: "Ice minus" involved the genetic engineering of a soil bacterium that produces a protein that provides a nucleating point for ice crystallization. Through a biotechnological intervention, the point at which soil would freeze had lowered. Although this was considered a desideratum by scientists interested in a longer growing season, it was popularly believed that such a bacterium would spread beyond test sites and wreak environmental havoc. (There was enough public response in this case that experimentation with this bacterium was ultimately stopped.)

Social Processes

Among the social trends triggered by both types of meta-technologies are the appearance of the network firm and related changes in organizational form, a renegotiation of relations among institutions and industries, reinforcement of socioeconomic class lines within and across societies, and several developments in the sociology of knowledge. This all began with the "informatization" (Nora & Minc, 1980) of society.

Informatization. It is the increase in the number of information technologies upon which we are dependent and the number of ways in which we are dependent upon them that led to use of the phrase "the information society." The process by which this has come about is described as informatization in parallel with the notion of industrialization, and the result of the process is an increase in the "information intensity" of technologies, organizations, and culture.

The industries that work with genetic information, beginning with agriculture, have, like other industries, become increasingly reliant upon the use of digital information technologies. They have also become more information intense as a result of the growing relative importance of biotechnological interventions. The use of new information technologies in agriculture is not new: The telephone was taken up early on by rural communities in both Europe and the United States to relieve social isolation and improve farmers' access to markets and prices. Indeed the highest levels of penetration of the telephone in the early 20th century in the United States was in the agricultural states of Iowa, Wisconsin, and Minnesota. Contemporary uses of digital information technologies include not only use of the internet to monitor markets, but also "precision farming" through the use of remote sensing, geographic information systems (GIS), and global positioning systems (GPS) to target the use of fertilizers, pesticides, herbicides, and water (Bowler, 1992; Haines & Joyce, 1987). If cost analyses are used to understand the nature of an undertaking, agriculture may soon be more accurately described as an information industry rather than a field activity (Flor, 1993).

Of course each new technology brings about structural change in the agriculture sector just as it has in other industries, encouraging coordination, standardization, and favoring those who can exercise economies of scale (Leeuwis, 1993; Phillips, 1989). There has been a deskilling effect because farmers must increasingly depend on off-farm decision making to determine how to treat their fields. The use of these technologies also reinforces the uniformity and chemical-intensive features of industrial agriculture and makes possible the "urbanization" of agriculture through hydroponics and cell cultures (Wilkinson, 1987).

The relative proportion of crops from genetically modified seeds keeps growing despite resistance on the part of consumers. A similar informatization of formerly industrial activities through the use of biotechnology is found in other fields; the bacterium at the heart of the critical Supreme Court case of *Diamond v Chakrabarty* (1980), for example, involved a genetically engineered organism designed to degrade components of crude oil to reduce the damage of spills.

Digital information technologies are also increasingly important to the effective use of biotechnologies. Government policy in support of biotechnology focuses on digital information technology support systems (de Freitas Filho et al., 2002). Bill Gates of Microsoft recently endowed a chair at the University of Washington in molecular biology devoted to the use of computers in genome analysis because increasingly, as discussed further in Lievrouw's chapter in this volume (chap. 6), mapping techniques are automated and genetic sequences are stored on disk (Boyle, 1996). Advances in the field of computerized measurement and control engineering have contributed to the development of bioreaction techniques (Roobeek, 1990). The computing needs of biotechnology are so great that Juno's Online Services' Virtual Supercomputing Network distributed computing initiative, which takes advantage of unused computing power on ISP subscriber computers, targeted biotechnology firms as a prime customer base (Eccles, 2001).

Networked Organizations. The "network firm" is so central to the operations of the contemporary economy that it is often referred to as a network, rather than an information, economy (Antonelli, 1992; Grabher, 1993). Although network firms do indeed characterize the information industries, and the use of digital technologies has made it possible for other types of organizations to similarly become transformed, the biotechnology industry led the way in experimentation with networked forms of organization—so much so that *Esquire* magazine once went so far as to describe the head of Genentech as the inventor of "post-industrial management" (quoted in Teitelman, 1989, p. 25). This has been manifested in at least three ways: the intertwining of small research shops with large transnational corporations, transformations in the internal structure of firms, and the blurring or loss of distinctions among the food, agricultural, chemical, and pharmaceutical industries.

Small research boutiques have been important to biotechnology since the early 20th century (Bud, 1993; Dechema, 1982). Over the last few decades, they have been critical to the industry as academics became involved with corporate startups to a unique degree. These firms are also unusual in that they are organized around the use of specific techniques rather than particular products (Krimsky, 1991). Universities, too, have become players, particularly since the Government Patent Policy Act of 1980 made it easier for universities to patent rights to discoveries that resulted from federally sponsored research. (There was a 300% increase in patent applications in the years immediately following passage of the Act [Krimsky, 1991].) However, although small firms are the source of scientific breakthroughs, it takes large corporations for the inventions that result to become logistically and economically feasible as innovations (OECD, 1988; Yoxen & Hyde, 1987). Large organizations prefer not to bring this expertise in-house both because the research and development (R&D) involved appear more likely to flourish in more intimate and nonhierarchical organizational environments, and because doing so lets corporations manage uncertainty through flexibility (Delaney, 1993). There are more collaborations between large corporations and external firms in biotechnology than in any other industry, with pharmaceutical firms often having dozens of collaborations taking place simultaneously (Powell, 1996).

Companies involved with biotechnology also led the way in experimentation with other characteristics of the network firm. Because the reputation of research organizations depends on their R&D prowess, many financial and managerial functions are contracted out (Powell, 1996). They use open organizational architectures that are fluid and only minimally hierarchical. Firms often have permeable boundaries with the bulk of their activity organized on a project team basis with groups from outside the firm, with linkages often formed through the efforts of legal and venture capital firms.

Patent strategies are often the basis on which partnerships are structured and business plans organized.

Shift in Institutional Relations. Digital information technologies have disrupted long-standing institutional relations by making it possible to cut out mediating organizations for many types of transactions, vastly increasing the amount of market information available and forcing institutions to reconsider not only which functions to retain internally, but even the products and services by which they are defined. The negotiations among universities, publishing houses, and libraries for management of various stages of the knowledge production, storage, and distribution functions is just one example of the ways in which institutional relations are shifting.

Similarly, the successes and aspirations of the biotechnology industry since World War II have affected relations among several types of pertinent institutions. In Schumpeterian terms, biotechnology is a competence-destroying innovation because it relies on types of knowledge different from those of the existing and mature pharmaceutical industry. Within academia in the United States, biotechnology raised the status of midwestern state universities relative to the private universities of the East Coast, as well as serving as a key interface between academia and industry (Bud, 1993). In the commercial world, biotechnology caused a shift in relations among the pharmaceutical industry, health care systems, and the government. Pharmaceutical company interest in biotechnology was spurred when it became clear that support for the drug industry from the federally funded health care system was declining (Powell, 1996). At the same time, Nixon's war on cancer resulted in the devotion of billions of dollars in research funds for cancer-related research, cementing the role of the federal government in directing the paths research would take and fixing bureaucratic relationships to the extent that by the late 1970s the government dominated biological research altogether. The combination of these factors explains the investment excitement that surrounded the 1970 discovery of the "oncogene," a gene believed to affect susceptibility to cancer once its DNA has been altered through viral infection (Teitelman, 1989). The expiration of major pharmaceutical patents in the 1990s provided further stimulus to the interest in biotechnology as a source of new, patentable products ("Waging Skyological Warfare," 1995). The ultimate result was a radical decentralization of the biomedical establishment, opening up new modes of financing, offering ways to get around established funding systems and stepping out from under the dominance of institutions like the National Institute of Health (NIH).

Class Divide. The digital divide—the phenomenon of a linkage between informational and socioeconomic class long referred to by sociologists as the knowledge gap—also has a parallel in the world of genetic information.

Within societies, there are differences in the kinds of food to which individuals have access and the size of organizations that will thrive. There has been a divide across societies since the moment Columbus launched the global flows of genetic information as a fundament of colonialism. The Green Revolution, the replacement of commodities with synthetic materials, and the current rush to mine traditional medical practices and the plants and organisms on which they depend for commercial purposes have reinforced tensions between the North and South.

It has always been those with the least resources who have had no choice regarding whether to eat foods developed through biotechnological interventions. Even several hundred years ago, as Braudel (1977) noted, it was the European poor who were forced to switch their diets to the new crops brought in from the western hemisphere—corn and potato. Although those in leisure classes were free to continue consuming foods for which there was a long-standing cultural preference, laborers were restricted to the new foods because those crops produced more calories per acre and thus helped support the Industrial Revolution. Today foods certified not to be genetically modified are more expensive, and thus are also not available to those without significant resources. Meanwhile the information intensity of agriculture and animal breeding now favors large industrial farmers because of the advantages of scale (Thorbecke, 1992), often resulting in a loss of jobs on farms as well as a forced change in lifestyle.

Biotechnology exemplifies the kind of asymmetrical relations between the North and South (Goldstein, 1988) that were of so much concern among those who promoted a New World Information Order (NWIO) in the 1970s and 1980s and who are promoting universal access to the Global Information Infrastructure (GII) today. The Green Revolution—the widespread diffusion of "improved" U.S.-bred seed throughout the developing world, largely supported by foundations and foreign aid—was intended to improve crop productivity and thus aid the economies and diets of recipient nations. The prediction offered by cultural anthropologist Carl Sauer in the 1940s (DeWalt, 1989), however, has come true: The introduction of improved seeds and industrialized farming practices destroyed the cultures, nutrition, and ecology of many societies. Although U.S. citizens may now have year-round access to fresh vegetables and fruits, for example, there has been no increase in caloric consumption, but rather less balanced nutrition and an increase in dependency in the countries from which those fruits and vegetables come (Roberts et al., 1986). Regional inequalities have in fact been exacerbated because, in the end, most Green Revolution projects benefited the rich but not the poor. Although corn hybrids briefly increased productivity, even that ultimately reversed and in some cases, as in South Asia, "improved" seeds for rice actually were counterproductive (Flor, 1993).

Because the Green Revolution transferred seeds, but not knowledge (Goldstein, 1988), its information intensity reduced the agricultural knowledge base through the loss of deep cultural knowledge regarding growing practices within specific ecological niches (McKibben, 1996). Even when contemporary techniques of biotechnology are introduced in developing societies, they generally remain in the hands of foreign nationals, and developing societies find they do not have the scientific or technical infrastructure to support activities on their own (Dechema, 1982). Often, in fact, the Green Revolution aided the information sector more than agriculture because so much information was involved in the course of knowledge acquisition, information generation, and institution building. Meanwhile the use of biotechnology has undermined developing country economies by replacing the commodities on which most of their economies are dependent—rubber, cocoa, sugar, vanilla, and other traditional crops—with synthetics (Clark & Juma, 1991; Goldstein, 1988).

The plunder by the agriculture-food-chemical-pharmaceutical industry of the genetic information of medicinal, cosmetic, and food plants used by traditional cultures—and the effort to do the same to the members of those societies as well (McNally & Wheale, 1996)—is currently the most serious North–South biotechnology issue. Conflict is so intense in this area that it has been described as "seed wars" (Longworth, 1992; Traill, 1988). The problem is clear: Claiming that the world's genetic resources are a commons, pharmaceutical and chemical companies based in the North collect germplasm from the developing world countries that have by far the greatest riches in terms of genetic diversity. Once gathered, that genetic information is processed via biotechnology, intellectual property rights are asserted over the products, and the medicines and other goods based on these products are then marketed at enormous cost to the same countries from which the original resources were taken. In response, a number of legal techniques are being developed, such as incorporating protection of genetic material and the transfer of knowledge and technology as conditions in contracts for the right to collect, expansion of intellectual property rights systems to cover indigenous knowledge, and growing support for the concept of a commons for pharmaceuticals. An additional problem leading to tensions between the North and South in this area is the practice of experimenting with the release of genetically modified organisms for which there is resistance in the North in regions of the South regardless of whether there is governmental approval, as the Pan American Health Organization and the U.S.-based Wistar Institute did when they conducted tests of a recombinant rabies vaccine in Argentina in 1986.

Ninety percent of the world's remaining biodiversity is in the South, particularly Asia and South America (McNally & Wheale, 1996). In the late

1980s, it was estimated that genetic information from the South had contributed $66 billion to the U.S. economy, and the figure would be even larger by now. That is more than the combined financial debt of Mexico and the Philippines, suggesting to some that a comparison of the North's "gene debt" to the South might be worth exploring (Mooney, 1988).

Sociology of Knowledge. Several of the chapters in this book touch on the impact of biotechnology on the sociology of knowledge, a matter also of wide discussion among those analyzing the effects of the use of digital information technologies. Trends in the sociology of knowledge include interdisciplinarity, a focus on innovation, the role of war in stimulating knowledge growth, the "dual use" nature of biotechnology and the knowledge it produces, and growing debate over postnormal science.

The role of biotechnology in repeated struggles for disciplinary standing, beginning in the 19th century but continuing today, has already been noted and is examined in more detail by Lievrouw (chap. 6, this volume). Biotechnology has long been interdisciplinary, providing a premiere exemplar of the value of such work in the extraordinary accomplishments of the physicists who turned from work on the nuclear bomb to questions about the nature of life after World War II (Fleming, 1968). For example, it was physicist Schrödinger who suggested in 1943 the metaphor of a "code script" that would explain control of every cell, drawing attention to ways in which Morse code could use as few as two signs arranged in groups of four to generate all that was needed to communicate, ultimately inspiring the discovery of the DNA helix.

Both biotechnology and digital information technologies create an environment in which there is a constant need for innovation. In agriculture, a varietal relay race is taking place requiring the constant insertion of new genes into elite cultivars to protect crops from vulnerabilities introduced by earlier innovations as well as to maintain a competitive edge (Kloppenburg & Kleinman, 1988). The value of both types of meta-technologies has increased the willingness of all countries to accept scientific innovation, and knowledge developed to support the diffusion of improved seeds during the Green Revolution (Wiegele, 1991) has been used to diffuse new information technologies.

War has been an influence on the direction and resources for research and development of both types of meta-technologies. Although industrial fermentation (zymotechnology) was seen as the miracle technology of 1900, it was not until these techniques were used to produce fuels and foods during World War I that the biotechnology industry really got its launch (Bud, 1993; Krimsky, 1991). The Cold War served as an impetus to biotechnology with its theory that the Green Revolution would be a way to prevent the spread of communism and Soviet influence. Vietnam offered an opportunity to experiment with herbicides and pesticides, which were regularly used in agriculture afterward. The U.S. Department of Defense declared in 1989 that biotechnology is a member of the class of technologies considered so critical to national defense that they are deserving of special funding and legal treatment by the government. This classification was made because biotechnology can produce disease-carrying agents and vaccines for those agents, dissemination protocols and manufacturing processes for toxic substances, equipment for the detection of biological warfare agents and protection against them, and equipment for containment and decontamination (Branscomb, 1993). Since 9/11 there has been a significant increase in the resources devoted to bioterrorism.

Post-9/11 biotechnology has also become a "dual use" science. The concept of "dual use" was developed for the purposes of designing export control regimes intended to prevent information and other technologies that could be used for either peaceful or nonpeaceful purposes from being exported to potentially unfriendly countries. Because knowledge about those technologies was also considered dual use, these regimes also applied to travel by individuals who possessed such knowledge and to access by foreign nationals to scholarly conference sessions on such topics. (Multilateral agreements putting such regimes in place include the Cold War Coordinating Committee for Multilateral Export Controls [CoCom] and the post-Cold War Wassenaar Arrangement on Export Controls for Conventional Arms and Dual-Use Goods and Technologies. At the national level, these agreements are translated into laws such as the U.S. Export in Arms Regulations [EAR] and the Information Technology Arms Regulations [ITAR].) In 2003, biotechnology and other biological sciences were essentially defined as dual use in the same way when dozens of the most important scientific journals, including the American Association for the Advancement of Science (AAAS) journal *Science*, announced they were inserting national security review into the peer review process as a defense against terrorism, and that several papers or parts of papers had already been withheld from publication on those grounds only (Black, 2003). This is yet one more reason that biotechnology is unfortunately also leading the way in the trend to reverse the norm and expectation that scientific results will be freely shared.

Finally, biotechnology has been both a site of discussion over postnormal science and a stimulus to that discussion. As have concerns over drilling for oil, standards for carcinogens, global warming, and the problem of disposing of nuclear waste, biotechnology presents a political challenge. All of these issues share the problem of assessing risks and benefits under conditions of uncertain knowledge, and for all of them scientific knowledge about their actual effects played only a limited role in public decision making, whether in the courts, Congress, or public agencies.

The Law

It is in the legal arena that the importance of these shared spaces for biotechnology and digital information technologies becomes clear. Their meta-technological nature has provided challenges to the legal process, particularly intellectual property rights issues. There has also been a reflection of their shared meta-technological characteristics in international relations and the law it produces. Finally, the convergence of biological organisms with digital technologies is leading to a growing role for technologies in decision making with structural effect for human society, a development that can be referred to as "posthuman law."

The Legal Process. The regulation and funding of both meta-technologies were motivated by governmental belief that each would be a solution to economic stagnation. For both the nation-state is a stakeholder because of the possibility of accruing benefits to government. Establishment of regulatory parameters such as intellectual property rights affects the development of both meta-technologies by altering the conditions in which they appear. Policymakers in both areas are faced with the problem of developing "transition policy" (Phillips, 1989) that is operable within the terms of legacy law, but also is responsive to qualitatively new conditions. Other shared features of the policy process include fragmentation, interdependence of law across levels of the social structure, the need for policy innovation, use of contracts for regulatory purposes, and a lack of knowledge on the part of policymakers about the meta-technologies being regulated.

It has been difficult to achieve coherent and successful policy programs for meta-technologies because each has multiple policy faces—industrial policy, social policy, economic policy, and so on. Policy venues are thus multiplied and treatment of single issues fragmented. Rabin (1987) identified at least 10 sets of stakeholders involved in biotechnology policy, each of which would like to lead the field: universities, national academies and professional societies, individual firms and trade associations, public interest groups, the executive branch of the federal government, Congress, state and local governments, the public, the press, and the legal profession. The biotechnology industry is subjected to oversight from at least seven federal agencies in the United States (Bailey, 1988), and the same complexity is found in the European Union (EU) (OECD, 1988). Efforts to develop a coherent approach often start with interagency efforts that result in deadlock (Krimsky, 1991). There can be tensions between policies designed to meet different goals within the same system (Duncan, 1989; OECD, 1988). There has often been confusion over the U.S. position on policies for meta-technologies because it is not clear what the policy objectives are or whether they will be consistent over time.

Although most law is still made at the national level, regulation of meta-technologies must ultimately be international. Decision making about bio-technology is found at four levels, each of which affects the others: international, national, local, and firm. Several issues are inescapably international in nature, such as concerns over biodiversity (Lueck, 1995) and use of bio-technology to expand agriculture into formerly hostile environments (Wilkinson, 1987). Interactions between foreign and domestic policy further stimulate policy interdependence; one of the strongest motives for U.S. policy in this area has been the desire to expand markets for U.S. products abroad (Mayer, 1986), and in Europe biotechnology has been included in aid packages for developing countries (Wilkinson, 1987). In some areas there are moves to harmonize the law, as in the 1988 call to establish international guidelines for the dissemination of new organisms as an outcome of the First International Conference on the Release of Genetically Engineered Microorganisms and the 1992 Biodiversity Convention.

Policy innovation has been required for the regulation of both types of meta-technologies, including development of new types of institutions and the use of demonstration projects (Duncan, 1989; Krimsky, 1991; Wiegele, 1991). One of the most vigorous areas of experimentation is the use of contracts to effect regulation. Klein (2003) and Zittrain (2003) explore this in the area of digital information technologies, and efforts to protect the intellectual property rights of indigenous peoples has stimulated the use of contracts for biotechnology regulatory purposes. For genetic information, contracts have the advantage that they do not require multinational legal agreements, and they avoid the legal difficulty of the difference between accidental and intentional discovery so important to intellectual property rights law (Cleveland et al., 1994; Sedjo & Simpson, 1995).

As with new information technologies, however, a lack of knowledge and sophistication about the subject matter on the part of policymakers is a barrier to successful policymaking for biotechnology. Quickly changing conditions and the wide range of organisms involved in biotechnology contribute to the knowledge problem (Mayer, 1986; Phillips, 1989; Rabin, 1987). An analysis of policymaking for research in this area in the late 1970s showed that it was driven by congressional appropriations subcommittees characterized by Hadwiger (1982) as an "anomaly of power lacking information" (p. 120). What he refers to as "low knowledge strategies" were the basis of decision making—legislators with expertise were only rarely invited onto pertinent committees, the seniority system provided long tenure on these committees which were thus slow to change, and committee decision making reflected an enduring resistance to consideration of ecological issues or the social and economic consequences of decision making. Knowledge generated out of scientific curiosity or for commercial gain does flow up to policy makers, although unevenly and unpredictably (Wiegele, 1991), but it is often difficult for them to understand how specific scientific knowledge relates to legal issues.

International organizations such as the World Health Organization (WHO), United Nations Educational, Scientific, and Cultural Organization (UNESCO), the Food and Agriculture Organization (FAO), and the United Nations Development Program (UNDP) have been important conduits for knowledge about biotechnology since the mid-1970s. For lack of an alternative, however, international organizations such as the International Development Bank (IDB) are applying management tools developed to analyze corporate investment strategies as a basis for biotechnology policymaking in the developing world (Clark & Juma, 1991). In 1988, the OECD specifically suggested that a first step for governments developing biotechnology policy should be development of agencies to serve as gateways to information. Technology assessment, where analysis of the problem should begin, is usually done sectorally, but in an area as complex as biotechnology needs to be done across the economy.

Intellectual Property Rights. Mutability of the subject and collective production confound efforts to assert intellectual property rights for the products of both digital information technologies and biotechnology. In some cases intellectual property rights issues for digital information technologies and biotechnology start from different types of questions, but present similar legal faces. Law developed to deal with one type of meta-technology then applies to the other, and there are parallels between legal approaches being developed for each. Patenting of process rather than product, extension of rights to collective producers, and licensing as an approach to protecting mutable information are particularly worthy of note. The impact of the law is significant for both meta-technologies because it establishes a hierarchy of preferred forms, with those that receive the highest levels of—or easiest—protection at the top. This can be exacerbated when intellectual property rights interact with other types of regulatory decisions, in the case of genetic information with the requirement for Food and Drug Administration (FDA) approval, and in the case of digital information with regulation of content, privacy, and security. These developments are also interesting from the perspective of the sociology of knowledge because patent litigation involving biotechnology today exemplifies the larger trend toward treating scientific research itself as the subject of legal dispute (Cambrosio et al., 1990).

Granting intellectual property rights to genetic information and the results of its processing has been problematic not only on ethical grounds,

but because it raises fundamental questions: Does biotechnology merely re-
arrange existing genetic materials, or does it produce new matter or a product of manufacture? How is it possible to grant intellectual property rights
for something that does not exist as a pure type, but only in multiple variants describable within sets of parameters? What is the difference between
discovery and invention? How can property rights be granted to individuals
or organizations when the knowledge their work is based on was developed
by farmers in various cultures over thousands of years? These issues are
often political. (Although not necessarily in the form in which the politics of
patenting genetic information was first introduced: When it was suggested
in 1901 that new types of potatoes ought to be worthy of patents, the notion
was rejected on grounds that it was "socialistic" [Kloppenburg, 1988].) Almost a million patents had been issued before the first one was granted for
a life form—to Pasteur, for purified yeast, in 1873 (Krimsky, 1991).

New forms of intellectual property rights are being developed for application to collaborative and traditional cultural forms stimulated by the realization that a "mining" of traditional cultures is necessary to realize the
commercial value of genetic information of use pharmaceutically (Greaves,
1994; King, 1994; Ruppert, 1994; Sedjo & Simpson, 1995). Of 119 drugs of
known chemical structure still extracted from plants in 1994 and in use,
over 74% were discovered by chemists trying to figure out how they worked
in traditional medicine (Laird, 1994). The typical computer software licensing agreement offers one model for indigenous peoples to adopt to be justly
compensated for the acquisition and use of the cultural knowledge essential to the use of genetic information in their environments, and to the types
of genetic information they have specifically developed. Licenses are useful
for applications of both types of meta-technologies when they involve inventions that are collectively developed and then enhanced and modified
during use. Typically such licenses are nontransferable, nonexclusive, and
perpetual, and they define the subjects being licensed comprehensively.
The license-granting entity—the software company, or the tribe—owns all
modifications regardless of form (Stephenson, 1994). The American Association for the Advancement of Science (AAAS) website archiving traditional
ecological knowledge also helps deter inappropriate patenting by docu-
menting prior art (Narayanan, 2002). Contracts are also being used for this
purpose.

Patenting activity involving both types of meta-technologies has moved
more and more into assertion of property rights over processes rather than
the products of processes. For digital information technologies, these are
actually known as "process patents," referring to patents that do not involve physical transformation such as those for software and business
methods (Kahin, 2003). In biotechnology, patenting a process is highly desirable because it provides the opportunity for income from everything

that happens upstream from that process. In 1995, for example, the NIH was granted patents over all ex vivo techniques used in gene therapy, meaning over all manipulations in which malfunctioning human cells are genetically altered outside the body and then replaced.

International Relations. Meta-technologies affect law between states as well as within them. The development of what is now called the International Telecommunications Union (ITU) in the 1860s to deal with cross-border use of the telegraph—which developed into a global system within two decades of coming into commercial use—has long been touted as the first international organization and a model of an international legal regime. Even before its establishment, however, an agreement had been put into place to manage flows of genetic information in the Codex in 1863, designed to increase world trade in food through mandatory international standards (Sklair, 1998). Now known as the Codex Alimentarius, this agreement deals with a number of issues important to those involved in biotechnology, such as treatment of microorganisms.

The impact of biotechnology on diverse arenas of international relations is discussed in more detail elsewhere (Braman, 2002a). In summary, the collection and redistribution of biological information was central to the imperial enterprise, and thus to the structuration processes shaping the global trade system. The transition to meta-technological treatment of genetic information, however, has destabilized the global structure as actors of a wide variety of types—many historically without any real power in either the international or, sometimes, local arenas—learn how to use the new forms of power that have become available. The results of this shift for international trade are threefold: There is an increase in turbulence and uncertainty in the international trade system, the focus of trade agreements has turned from product to process, and new modes of action and types of agency have been enabled by the new meta-technologies. Biological and digital information technologies are now converging in the area of international trade not only in what is being traded, but also in the revamping of the intellectual property rights system because changes directed at problems raised by either are applied to both.

Genetic information and biotechnology have also been important throughout human history both as causes of war and as tools used in the fighting of war. The ways in which this has been so are not always obvious; the impact of improved techniques for food preservation on military logistics was enormous, but was generally understood—or, even, visible—only to specialists (Macksey, 1989). Deep linkages between genetic information and national and cultural identities has provided an additional symbolic dimension to the role of biotechnology in war historically. Today tensions over whether to accept genetically modified foods increase the difficulty of reaching a consensus between nations also in conflict over other matters. The military situation is less stable than it was during the Cold War in part because of the new possibilities of information warfare in both digital and bioterrorism forms. Meanwhile one cutting edge of U.S. military research and development calls for work that will "codesign" soldiers and weaponries so that soldiers of the future may be best fit to work with the digital information technologies to come.

The effects of biotechnology on agriculture in international relations in recent years have been several. They have changed the rules of the game regarding the international division of labor, pulling out from under many former colonial societies the economic underpinnings on which they had come to rely. In turn new opportunities for endogenously driven change in domestic agricultural practices may be opening up with increasing biotechnological sophistication. Acceptance or rejection of genetically modified foods has become a negotiating tool for nation-states seeking to improve conditions for their own agricultural producers in the global market or for those trying to protect certain cultural, health, or environmental positions.

Posthuman Law. A more detailed analysis of the ways in which the convergence of biological and digital information is yielding new relations between the human and machinic worlds is also offered elsewhere (Braman, 2002b). Although it has been an unspoken assumption that the law is made by humans for humans, that assumption no longer holds. The subject of information policy is increasingly flows of information between machines, machinic rather than social values play ever-more important roles in decision making, and information policy for human society is being supplemented, supplanted, and superseded by machinic decision making. The use of biological models for computing and networking and the appearance of artificial life forms capable of autonomous agency and self-evolution—important manifestations of the convergence between the two types of meta-technologies being discussed here—are key to and exacerbate these trends. As the barrier between the human and machinic continues to fall with implantation of chips within the body and other types of intimate relationships, new types of regulatory tools are also becoming available to the legal system based on the convergence of the meta-technologies of genetic and digital information.

Convergence of Genetic and Digital Information

The importance of the shared spaces described earlier will grow as the convergence of the organic and digital worlds becomes more pervasively realized. A few of the places we are seeing this trend include the use of biological models for computing, computing with biological organisms, artificial life, and the cyborg.

Beginning with the study of neural networks for models of possible computing processes, biological models have been important to the design of hardware and software. Today's parallel architectures process multiple levels of information simultaneously in ways analogous to those of biological structures, appearing more intuitive than programmed, more sensitive to environmental data than earlier types of computers, capable of learning from experience, and able to filter data in ways that approach the scope of "wet systems" like human cognition. Programming spreads ubiquitously through computing networks, like enzymes waiting to be triggered into action by an informational prompt or environmental cue (Hookway, 1999). Intelligent agents increasingly exercise agency—even autonomously—as well as reproduce and modify themselves (Turkle, 1995).

There has also been a reliance on biological metaphors from the beginning of emergent artificial intelligence (AI) research (Turkle, 1995). A mixture of centralized control and bottom–up evolution characterizes a class of programs developed by John Holland in the 1960s that received the name of genetic algorithms, meaning strings of randomly generated zeros and ones defined as a chromosome and used to search problem space for possible solutions to any given task. Genetic algorithms can be used for tasks such as evolving increasingly efficient computer programs. However, such algorithms do not always work like Darwinian evolution, through natural selection. In fact, when programmers simply forgot to simulate evolution through mutation at all, they were surprised to find that they still got results that looked a lot like evolution. As a result the use of genetic algorithms has resulted in discovery of evolutionary processes previously unfamiliar to biologists, but that have in turn influenced biological thought (e.g., "crossover," the mixing and matching of two gene strands, appears to be at least as important a motor of evolution as mutation). Artificial life has the advantage that it is possible to look at the equivalent of hundreds of thousands of years of evolution in short periods of time. Although early experiments were limited by user determination of fitness, since the mid-1980s the most important work with artificial life involves neither any predetermined goal for evolution nor a breeder—just interactions within and responses to digital environments.

With the ability to create electronics at the nano-scale, it has also become possible to use genetic information for computing. Nano-electronics using genes from single-celled organisms now form the basis of computing structures 10 to 100 times smaller than today's electronic components ("NASA Breakthrough," 2002). DNA computing uses strands of DNA to process information with biological molecules and, depending on the problem, can be more than 1 billion times faster than existing computers. It is believed that with this approach it might be possible to create databases far larger than that of the human brain ("A Thousand Billion," 1995).

Folks disagree about the moment at which the cyborg, or the post-human, began to appear. Some point to the synthesis of urea in 1838, which stimulated efforts to synthesize natural products in the laboratory, as the moment at which the distinction between natural and artificial disappeared (Bud, 1993). Dewdney (1998) believes it is the first computer generation, born in the 1980s, with its attraction to transformative toys. There are also differences in just what is meant by this transition: For Baudrillard (1983), it is absorption in simulation that makes one a cyborg; for Beaumont (1984), it is the merging of the individual with machinery; for Stone (1995), it is the re-definition of personality for the technosocial environment; for Fukuyama (2002), it is the expectation that the accumulation of genetic interventions will transform the human into another species; and for Naficy (1996), to be transhuman is simply to be an evolutionary human being.

However it is defined, the posthuman, transhuman, or cyborg world will place different demands on individuals and societies. Cetina (1997), who prefers the term "postsocial," suggests the transition will be an extension of the trend toward the definition of identity in relation to the objects in one's environment, rather than to the social or spiritual realms. The impact on subjectivity will be unavoidable, leaving some optimistic (Guattari [1995] argues that the means of production of subjectivity will be appropriated) and others pessimistic (Lessig [1999] suggests that the sense of the self will be irrevocably damaged by being subjected to constant surveillance). Der Derian (1990) locates the cyborg at the level of the entire social system as well as in the individual, as did Virilio, who saw the human body as the site in which global technological and capital logics are played out (Beller, 1996; Wilson & Virilio, 1996). The individual is not entirely passive in the transition to the posthuman world. As Stone (1995) notes, one of the motivations is "cyborg envy," the desire to penetrate an interface and merge with a technological system. Artist Eduardo Kac and others have gone so far as to embed electronic chips in their bodies (Dickey, 2001).

META-TECHNOLOGIES AND POWER

All of this discussion about the shared features and spaces of meta-technologies matters at root because of its implications for the nature of power. The concepts of "biopower" and related social features such as "biomodernism," the "biosocial," and the "biotechnic" are found in thinkers like Virilio, Baudrillard, Rabinow, Foucault, Barthes, Mumford, and Heidegger (Kroker, 1992; Stone, 1995). Formulations are quite different, but all share appreciation of the way in which power is increasingly exercised over persons specifically insofar as they are thought of as living organisms, as exemplified in the work of Foucault, and of the way in which knowledge about the transformation of human life is now being brought explicitly into power calculations, as exemplified in the work of Rabinow. In each case analyses link genetic information and digital information as revelatory of shared features from the micro- to the macro-level of social processes. Within the field of communication, such concepts are important to explorations of organic effect of digitally generated simulation (Baudrillard, 1983), as a way to think about the technosocial nature of emergent forms in cyberspace (A. Stone, 1995), and in examination of the use of organic and mechanical metaphors of human systems (see Ritchie, chap. 2, and Wildman, chap. 3; Haraway, 1976; Peters, 1988; J. Stone, 1991; Woodward, 1994).

Political scientists have long focused attention on power in three forms:

instrumental—power that shapes human behaviors by manipulating the material world via physical force;

symbolic—power that shapes human behaviors by manipulating the material, social, and symbolic worlds via ideas, words, and images; and

structural—power that shapes human behaviors by manipulating the social and material worlds via rules and institutions.

In today's information-intense society, however, it has become clear that information is not only a distinct form of power in its own right, but has moved to the center of the stage, dominating the uses of all other forms of power and changing how other forms of power come into being and are exercised. The terms "genetic" or "informational" can be used to describe this form of power as it appears at the genesis, the informational origins, of the materials, social structures, and symbols that are the stuff of power in its other forms. In doing so it simultaneously extends power over the noetic universe as well. It can be added to the prior typology this way:

genetic or informational—power that shapes human behaviors by manipulating the informational bases of the material, symbolic, and social worlds.

Genetic power is a particularly important form of power today because it is that which takes the greatest advantage of the distinct characteristics of this stage of the information society, the harmonization of systems—of nationally based information and communication systems across geopolitical boundaries, of different types of information and communication systems with each other, and of information and communication systems with other types of social systems (Braman, 1993, 1995). In such an environment, information flows have structural effects as powerful as those traditionally associated with the law. As a result, the ability to shape those flows is the most important form of power of all. This is the type of power Lessig (1999) talks about in hypothetical U.S.-based detail in his popular *Code and Other Laws of Cyberspace*, Dezalay and Garth (1998) analyze in its international negotiational expressions, and Lewis Branscomb (1993) refers to when he notes that both digital information technologies and biotechnologies are of strategic concern.

Whether it is expressed in discourse, social practices and institutions, cultural habits and frames, economic relationships, laws and regulations, or the literal converging of biotechnology and digital information technologies and the information processed by each, it is the impact on the nature and manifestations of power that should guide our future research agenda in this area.

PART

II

THE CONCEPT
OF INFORMATION

2

Information as Metaphor: Biology and Communication

David Ritchie
Portland State University

Information has provided a powerful metaphor for explaining the complexities of evolution and its complementary engines, DNA and environment; some of the theorems of information theory (Shannon, 1949) illuminate the constraints on the intergenerational stability of the genome. Dennett (1995) develops the information metaphor into a nontechnical explanation of the fundamental concepts of evolution of biological species, which he then extends to evolution of the ideas, practices, and patterns that constitute culture. However, Oyama (2000) claims that the application of information theory to evolution has sustained misleading dualisms between environment and heredity, material and form. As a communication scholar, one is struck by the parallels to the ambiguous trace left by information theory on our understanding of symbolic communication between and among humans. In both cases information theory has literal and figurative applications to the phenomena of interest, and the conflation of a literal application of the mathematics of information theory with the underlying metaphoric understanding of certain complex phenomena in terms of signal transmission can lead the unwary into errors that obscure as much as they clarify.

In Ritchie (2003) I show how the mathematical concept of statistical probability has become conflated with an epistemological concept of logical probability through a complex series of metaphorical mappings, and I use metaphor analysis to tease these concepts apart and explicate their relationship. In Ritchie (2001), I show how the interpretation of metaphor often requires a consideration of the broader sociocultural context as well as of the immediate textual context: In some cases expanding the context within which a metaphor is interpreted leads to a radical reformulation of meaning. In this chapter, I propose to apply a similar analytic technique to teasing apart the literal and metaphorical uses of information and information theory in writings about biology, and show how the conflation of the literal and metaphorical in writings about biology parallel the conflation of literal and metaphorical understanding of information theory in writings about human communication. I begin with a brief review of Shannon's basic ideas, including his overt use of certain compelling metaphors as a way to explain abstract mathematical ideas. I then summarize Dennett's (1995) use of information theory to explain evolution theory and Oyama's (2000) critique of the use of information theory in genetics. Finally, I return to an assessment of the interplay between the literal and metaphorical applications of information theory.

METAPHOR

Lakoff and Johnson (1980) argue that conceptual thought is fundamentally metaphorical. They have shown how our understanding of even the most basic conditions of our existence can be traced to metaphors grounded in our experience of the physical conditions of life. For example, they identify a family of orientational metaphors ("The stock market is up ten points," "I'm feeling low today," "I felt a rising sense of anger and shame") and show how these metaphors link abstract ideas to embodied experiences such as sleep and waking, illness and health, emptiness and fullness. Lakoff and Nunez (2000) extend this reasoning in an ambitious attempt to show how even the most abstract forms of mathematical reasoning develop as a chain of metaphoric extensions from a small handful of innate perceptual abilities.

Vervaeke and Kennedy (1996) criticize the concept of implicit metaphors, contending that the identification of root metaphors in the interpretive process is indeterminate, and that root metaphors are implicit—not in the sense of inevitability or uniqueness, but rather "in the sense of 'waiting' or 'available' or 'apt once they are mentioned' " (p. 227). According to Vervaeke and Kennedy, metaphors do not necessarily constitute thought, but are chosen from an indeterminate range of possibilities to express literal ideas. A set of metaphoric statements may be consistent with any of a number of root metaphors, each with its own distinct set of entailments, and the selection of any particular root metaphor for analysis will almost always be underdetermined. Thus, multiple interpretations of the same body of expressions are often plausible, and metaphors may be chosen deliberately to express particularly complex or subtle ideas.

Vervaeke and Kennedy (1996) also point out that novel metaphors can be produced by extending an identified root metaphor, but the novel metaphors thus produced "may or may not be apt or may run counter to the examples in the original corpus" (p. 227). Not all the entailments of a root metaphor can necessarily be mapped onto the target concept. As a consequence, even an interesting and informative metaphor may fail to generalize, and metaphor analysis is unlikely to lead to certainty about underlying meanings.

I have elsewhere (Ritchie, 2002) extended Vervaeke and Kennedy's argument by suggesting that the analyst may usefully consider more than one root metaphor while trying to make sense of discourse. Metaphor analysis can help unpack implications of the language used in discussion, and thus broaden our understanding of underlying concepts. The analyst must carefully consider entailments of a proposed root metaphor, rejecting those that seem contradictory with the intent of those who use the metaphor as well as those that simply do not fit and accepting only those that convincingly contribute to understanding the target concept.

Kittay (1987) mentions a set of expressions that may be taken either literally or metaphorically. For example, Kittay cites a prisoner explaining why he did not escape by saying, "My hands were tied. I could do nothing." Or we might say of an overweight friend who has chanced into a sinecure, "Fred is fat and happy," or laughingly say of a grossly exuberant friend, "Tom is an animal." Whether the intention is literal or metaphorical can be determined only by the context. In Ritchie (2001), I extend this idea further, showing how the interpretation of metaphorical/literal expressions may depend not merely on the immediate context, but on the extended context that includes social, cultural, and political assumptions and beliefs. For example, Lakoff and Turner (1989) analyze common euphemisms for death, such as "passed away," "was taken from us," in terms of root metaphors, "LIFE IS BEING HERE" and "DEATH IS DEPARTURE." But many religious traditions identify the person with an immaterial soul, distinct from the body, and regard death as literally both a departure and liberation. Moreover, for a person familiar with such a religious tradition (such as a person raised in but no longer belonging to a conservative religious group), the usage may be metaphorical, but based on the root metaphor, "DEATH IS THE DEPARTURE OF AN IMMATERIAL SOUL." The use of one of these expressions cannot be understood without considering the cultural, religious, and sometimes even political contexts.

In this chapter, I extend this line of inquiry further by showing how the proposed approach to metaphor analysis can help resolve ambiguities resulting from a conflation of literal with metaphoric entailments of complex ideas such as information. Information, information theory, and several related concepts are often used in a way that, like "fat and happy," "animal," or "passed away," can be understood as literal, metaphorical, or both at once depending on the context. A further consequence of Vervaeke and Kennedy's argument is that both the rhetor and the rhetorical analyst must consider metaphors and the relationship between metaphoric and literal meanings in a deliberate way. Primary tasks of metaphor analysis when such complex metaphorical usages are involved include opposing the conflation of the metaphorical with the literal by first teasing apart the metaphorical and literal meanings and then separating apt entailments of the root metaphor(s) from inapt entailments.

INFORMATION THEORY

Information theory has been understood with reference to several distinct aspects of Shannon's work as well as a broad array of more or less related work. As developed by Shannon (1948), information theory began as an attempt to describe the constraints on transmitting signals by way of a medium (notably an electronic circuit) of known capacity. In this sense, *information theory* refers to Shannon's theorems stating the relationship among channel capacity, transmission rate, and accuracy. More specifically, people have sometimes identified information theory with Shannon's expression for the variance or "scatter" within a set of elements such as a code or alphabet, in terms of binary digits or "bits," $H = -k\Sigma p(i)\log_2 p(i)$. This formula for H, which is virtually identical with the formula for entropy, plays a central role in Shannon's development of the theorems (discussed later) and thus lies at the heart of information theory. It has also come to serve, especially in the field of Communication Studies, as a powerful icon that seems to guarantee prestige and scientific status to any text in which it appears (see Ritchie, 1991; Zencey, 1991).

Sometimes Shannon's famous model of a transmission circuit (source–message–receiver) is understood as information theory. Recently, the much more general theory of digital computation (information processing), made possible in part by Shannon's original formulation, has also been taken to constitute information theory. In general, any discussion of factors affecting the organization, description, transmission, or replication of patterns, such as the various formulations of chaos theory and self-organizing complexity, may also present claims for inclusion within the general meaning of *information theory*. Here I focus primarily on the model of a signal transmission system and the theorems that describe the constraints on transmission through such a system.

It is worth recalling the problem Shannon originally addresses, which was to develop a theoretical basis for increasing the efficiency of electronic circuits. Although Shannon was an employee of Bell Labs, it is clear throughout his article that he intended the theory to have application well beyond telephone circuits, including radio and TV broadcasts and indeed any medium of signal transmission whatsoever. Indeed by titling his article "The Mathematical Theory of Communication" and by drawing illustrative examples from such diverse activities as cryptography, the structure of language, and crossword puzzles, Shannon explicitly encourages a broad application of his work.

Electronic signals are subject to random perturbations, metaphorically referred to as *noise* because of the way human beings subjectively experience the sounds produced in a telephone or radio receiver by such perturbations. The second law of thermodynamics, famous as the "law of entropy," implies that such random perturbations are both unavoidable and unpredictable because electronic signals are a highly organized form of energy, and energy tends to degrade from a more ordered to a less ordered form. However, the effects of these random perturbations can be detected and counteracted through various forms of redundancy—for example, by repeating the entire message or inserting check digits at the end of each string within a message. Part of the question for Shannon, then, was to formulate measures for the maximum capacity of a circuit, the degree to which random perturbations reduce that capacity, and the limits to the improvement of accuracy and transmission rate that can be achieved by introducing redundancy into the code.

Any message can be transmitted through any medium by encoding it into a series of detectable variations. For example, the telegraph relies on patterned interruptions in an electrical current; the telephone and radio encode the analog qualities of the human voice into analog electromagnetic waves. (Digital transmission encodes the voice into sequences of digits that characterize the amplitude of the sonic wave at regular intervals several times per second.) Semaphores create patterns of flags; Paul Revere's famous code used a pattern of lighted lanterns. The human voice encodes words into patterns of sound waves; alphabetic writing further encodes each discrete sound into patterns of letters.

In the simplest form of medium, such as a telegraph wire, Paul Revere's lanterns, or the bonfires by which news of the fall of Troy was spread, the code is digital. Familiar contemporary examples of digital codes include Morse code and the three taps on the ceiling or two bangs on a radiator pipe in the old country-western song (Levine & Brown, 1970). Reflection on the nature of messages leads to the realization that they always take a form very much like a formal code. In particular, there must be variation (if everyone went around humming middle C, there would be no such thing as language), but there must also be organization (if everyone went around making a perfectly random series of sounds, there would be no such thing as language). The set of distinguishable elements, along with the rules by which they are matched with elements from any other set and assembled into messages and the rules specifying the internal structure of allowable messages, constitutes the code.

Digital codes such as Morse code suggested a ready metric for the capacity of any code: How many binary digits (e.g., zeros and ones, on and off states, bangs on the radiator pipe) are required to specify uniquely any particular element in the code? This problem is metaphorically similar to the problem in thermodynamics of uniquely specifying a particular energy state of a system (e.g., the dispersion of molecules in a container of gas). That is to say, the problem of measuring the capacity of a code is metaphorically related to the problem of measuring the entropy of a self-contained physical system. Gibbs' formula for the entropy of a physical system provided a convenient and intuitively apt expression for the amount of variation among elements of a set, so Shannon adapted it to his own needs. The first part of his paper (Shannon, 1948) is devoted to showing that a modified version of Gibbs' formula (H, given earlier) is suitable to the task. Indeed so suitable is it that it is difficult to imagine a better measure. However, Shannon explicitly refrained from claiming that H is a unique or even the best possible measure. He simply shows that it is adequate to the task.

When Shannon uses the word *entropy*, as in "entropy of the signal" or "entropy of the source," he always refers to the degree to which elements of the source or signal are scattered. The value of H, as a measure of information capacity of a code or transmission channel or as a measure of the amount of channel capacity required to transmit a certain message, reaches its maximum when the distribution of elements (e.g., dashes and dots, letters of the alphabet) is perfectly random (i.e., when at any point in a transmission each element is equally probable). However, a perfectly random distribution of elements has the undesirable characteristic that, if a random event (i.e., literal physical entropy) changes one element of the code, the error is undetectable and consequently uncorrectable. Error detection and correction can only be accomplished if redundancy has been introduced into the code. Redundancy can take various forms; the simplest form is merely to repeat every message. For example, military orders or reports are almost always repeated verbatim. The constraints imposed by every natural language on spelling and syntax also increase the possibility for error detection and correction at the expense of increased redundancy and consequent reduced channel capacity. (Another undesirable characteristic of a maximum capacity code is illustrated by the Levine and Brown [1970] song mentioned earlier: If every possible sequence of bangs on the radiator pipe is used with equal frequency in the code, it would be impossible to distinguish an intentional message from simple malfunctioning of the heating system.)

Shannon's theorems are based on comparing the "entropy of the source," the statistical dispersion of elements in the source, with the "entropy of the channel," the statistical dispersion of elements in the code. Throughout most of Shannon's discussion, *source* refers to the set of elements of which a message may be composed. At the simplest, for transmission systems based on written English, the source is the letters of the alphabet. The maximum value of H—that is to say, the maximum capacity of the alphabet as a code—would be achieved if every letter appeared with equal frequency at every point in any message; in other words, if the assembly of a message were unconstrained by rules of spelling and syntax. But in fact, some letters (such as e, a, and s) appear much more frequently than others (such as x, z, and q). Moreover, if a message includes the string "th," the next letter is very likely to be a vowel and will virtually never be "z" or "n." Expand the preceding string to fewer than a dozen letters and the next letter is often all but certain. Readers make automatic and usually subconscious use of this redundancy: When we come across a phrase such as "The chances are quite thn," we insert an "i," reading the phrase as "The chances are quite thin." Because of the redundancy built into the language, nothing else makes sense. Often the error correction happens subliminally—a fact that can pose a great challenge to proofreaders who may see the word as *thin*, with no awareness that the visual circuits in their brains have filled in a missing "i."

It is possible, by studying the way letters are combined into words and words into phrases within a given language community, to detect patterns such as these, which are repeated with sufficient frequency that channel capacity can be saved by encoding each of these patterns as a unit. Thus, when developing a code for transmitting news stories, entire stock phrases (stereotypes) may be treated as a unit and assigned a single sequence of digits in the transmission code. A code might even be designed to permit maximum efficiency in transmitting messages from a single speaker or author (e.g., the president of the United States). So the source can refer to the collection of messages typically sent by a language community or even by a particular writer or speaker. Analysts of literary style sometimes use this technique, for example, to identify the author of a disputed text by comparing the conditional frequencies of elements in the text with conditional frequencies of the same elements in authors' known works. The important thing to realize is that *source*, as used in Shannon's theorems, is distinct from the use of the word *source* as in "the source of a quote" or "a newsworthy source." When information theory is applied to genetics, *source* in this sense may refer to the collection of four nucleic acids or the collection of genes (patterned sequences of the four nucleic acids) that appear in a particular genome. The chromosomes of one particular parent then constitute a message composed of units drawn from that source.

Note that the relationship of H as a measure of transmission capacity to entropy, like the relationship of signal perturbations to noise, is primarily metaphorical even as it is based on a literal linkage. Entropy literally affects signal transmission inasmuch as signals are always transmitted in the form of organized energy states in a self-contained physical system, and these organized energy states are always subject to degradation through random physical events. (For example, the pattern of ink on this page is subject to alteration by the passage of ions through the atmosphere, by the abrasion of readers' fingers, and by spills of coffee, Coke, or wine on the page.) However, codes and messages are abstract entities distinct from the physical systems described by the laws of thermodynamics: Describing the former in terms of the latter is unavoidably metaphorical. Messages must be stored and transmitted by way of patterns in a physical medium, but the patterns are distinct from the medium in which they are expressed. If the medium is altered in some way (e.g., if lightning strikes a telegraph wire), the observed pattern may be altered as a result, and the message may be partially altered or lost. H, as a measure of the *entropy* of a source or message, refers to the degree of scatter or apparent randomness in the distribution of elements within the source or message. As a measure of the *"entropy"* of a channel, H refers to the degree of scatter or apparent randomness in the distribution of elements in the code by which messages are transmitted. H has only a metaphorical relationship to physical entropy—the tendency, articulated in the Second Law of Thermodynamics, of all energy systems to become more disorganized. Literal physical entropy affects the transmission system by creating the randomizing noise that introduces errors into transmitted messages. Thus, as a measure of the noise in a channel, H refers to the probability that any one element of a transmitted message will be altered because of random events affecting the transmission.

According to Zencey (1991), entropy has become a widespread and popular metaphor for the relationship between order and disorder often applied in areas such as history, morality, and even religion, in which thermodynamics has little if any apparent relevance. It is true that all of these events involve physical entities in some way, and all physical entities are subject to the laws of thermodynamics. The random events described by the laws of thermodynamics could conceivably lead over time to the breakdown of religion and morality—but entropy in this literal, physical sense rarely plays any serious role in theories of religion, morality, or statecraft. It is rare that enthusiasts for the concept of entropy take the trouble to show how physical entropy might be related, causally or otherwise, to the examples of disorder that concern them: The usage is virtually always metaphorical. Not surprisingly, a similar fate has befallen information theory and the concepts associated with it, including information, noise, and redundancy (Ritchie, 1986, 1991).

Unfortunately, metaphorical abuse of information theory was licensed almost from the outset by Weaver's (1949) interpretive extension of the theory (see Ritchie, 1986, 1991, for detailed discussion and critique) and by the advice given by prestigious scholars such as Wilbur Schramm (1955) to "read Weaver first." It did not help that many students and scholars of human communication have lacked the mathematical sophistication to understand Shannon's original formulations and so never got around to reading Shannon at all. Consequently, information, like entropy, mutated into a quasimystical concept empowered by the aura of apparently impenetrable mathematics (see also Zencey, 1991). Indeed the two concepts are frequently linked—for example when information is equated with negative entropy and credited with the power to neutralize the unpleasant consequences of the second law of thermodynamics, such as old age and even the eventual descent of the universe into total disorder. (The right information could, of course, be used to counteract entropy and slow or even reverse the aging process within the bounds of a single organism, but it would be at the cost of a much greater increase in the entropy of the surrounding environment.)

It is useful to remember how much Shannon draws on other metaphors in developing, explaining, and justifying his ideas. He explains the problem of specifying a particular message out of a group of potential messages and related ideas such as redundancy in terms of several more or less familiar processes including cryptography, linguistics, and crossword puzzles. The word *information* is metaphorical—a metaphor reinforced by Shannon's extensive use of everyday human communication activities such as crossword puzzles (and not so everyday activities such as cryptography) to illustrate his concepts.

"Inform" comes from the Latin verb *informare*, to inform or give form to; *informare* is derived from *in-* plus *forma*, form (*Oxford English Dictionary*). Most directly, then, *information* is that which informs or gives form to, or the process of informing or giving form to. The information capacity of a circuit or code (what H measures) can be understood metaphorically as the capacity of messages formulated in the code and transmitted via the circuit to create form (through the decoding process). What is "informed" is unspecified: Shannon (1949) explicitly disavows any interest in the meaning of the messages transmitted (the "forms" resulting from the messages), and later (1956) expresses considerable concern over the casual metaphorical extension of his ideas to the realm of human communication and to meanings.

In the study of human communication, information theory has proved to be a singularly dry well, at best evocative (Finn & Roberts, 1984), at worst downright misleading (Ritchie, 1986). The real news of Shannon's work, which has been largely ignored by scholars of human communication, is in his theorems that specify the relationships among the complexity of a code, the capacity of a channel, the degree to which the channel is subject to random perturbations (noise), and the amount of redundancy built into the code. Communication is always based on transmission of some sort of signal, whether it be spoken or written language, a nonverbal gesture such as a touch or nod, or, perhaps, brain-waves or "psi" energy. Because transmission of any sort of signal is subject to random degradation, any communication channel whatsoever is a "channel with noise."

Based on these considerations, Shannon demonstrates that (a) perfect, error-free transmission is unachievable; (b) at least under some circumstances transmission can be brought arbitrarily close to perfection; but (c) incremental decreases in transmission errors are costly in terms of increased channel capacity, decreased rate of transmission, or both. To see why this is so, consider again the military rule that every order must be repeated: There is a finite, if small, probability that a randomizing process will produce exactly the same error in both versions. Another form of redundancy is the use of a "check digit," representing the sum of all digits in a sequence of specified length. If one random event alters one of the digits in a string, another random event can alter the check digit in such a way that it appears to add up. In general, no matter how cleverly the code is constructed, no matter how many error-detection features are built into the code, it is always possible that some sequence of random events can both produce an error and alter the error-detection part of the code so as to produce the appearance of an error-free transmission. Thus, in a universe in which random events occur, error-free transmission is impossible. However, it is possible, by introducing additional error-detection features into a code, to create a situation in which a huge number of improbable coincidences must occur for an error to slip by undetected. Thus, it is possible, through added redundancy, to reduce the rate of actual errors until it is arbitrarily near zero (but never equal to zero). Finally, every form of redundancy requires its own part of the signal, and thus reduces the amount of the signal left for transmitting the message. Each increase in accuracy of transmission carries the cost of a decreased rate of transmission.

With respect to human communication, it follows that if perfect transmission of signals is unachievable, perfect intersubjectivity is unachievable. Any theory that relies on more than an approximate intersubjectivity is fatally flawed. To most observers of human communication, it probably comes as no news that perfect intersubjectivity is unachievable. But had theorists taken seriously the true content of Shannon's (1949) theory, that is to say the theorems, a long and fruitless theoretical debate over intersubjectivity might have been avoided.

Incidentally, similar reasoning applies to the research process. A researcher observing a phenomenon of any sort is detecting signals: Both the signals and the process of detecting the signals are subject to random perturbations. Although it is in concept possible, through instrumentation and methodological improvements, to come arbitrarily close to perfect reception, absolute perfection can never be achieved. Thus, true certainty is unachievable.

However, in both instances, it must be noted that the major sources of erroneous understanding are in the processes of interpretation, not in reception. Pending major breakthroughs in brain science, information theory can only be applied metaphorically to the processes of interpretation. To return to the example of military orders, the greatest source of error in orders is faulty reasoning on the part of the officer giving the orders or preconceptions on the part of the person receiving the order. In the first place, if a bad order is given once, it will almost certainly be given the second time. In the second place, if preconceived expectations as to what the superior officer is likely to command leads the subordinate to mishear or misinterpret an order, no amount of redundancy can ensure that the mistake is corrected. The movie *Crimson Tide* (Scott, 1995) illustrates the problem of interpretation quite well, when the overtrained captain misinterprets a partially received "stand-down" order as mere noise and nearly starts a nuclear war despite the interpretive redundancy built into the dual-command structure.

Although information theory describes the absolute limits to intersubjectivity, on the one hand, and to scientific discovery, on the other, the greater uncertainties associated with interpretive processes keep us so far inside those limits that information theory is unlikely to lead to any truly useful insights. Application to human communication is based on an implicit metaphor, "discourse as signal transmission," and application to epistemology is based on an implicit metaphor, "nature as transmitter." Both metaphors require careful inspection and justification before application of Shannon's (1949) theorems can be accepted. Signal transmission is certainly part of discourse, which requires transmission of messages in some physical medium, but it is commonly a very small and ordinarily the least significant part.

INFORMATION THEORY AND EVOLUTION

Although other writers have applied information theory to biology, in both literal and metaphorical senses, Dennett's (1995) account of evolution theory provides a particularly useful example. Dennett sets out to generalize the ideas embodied in evolution theory to other branches of knowledge, in particular to human culture. Dennett develops a concept of *design space* based on Argentine poet Jorge Luis Borges' (1962) account of the "Library of Babel." The Library of Babel contains all the information in the universe, which is for computational (and bibliographical) convenience organized into books of 500 pages each, with 40 lines of 50 spaces on each page. Each space is printed with 1 of 100 possible characters (including spaces, punctuation marks, etc.). If every possible book is included in the Library of Babel, there will be $100^{1,000,000}$ books in the library, including, for example, 100,000,000 versions of *Moby Dick* that differ from the canonical version by a single typographical error. The Library of Mendel, consisting of every possible combination of the four nucleic acids, forms a subset of the Library of Babel.

There are in the Library of Babel an uncountably large number of brilliant novels and poems, most of which will never be written, an uncountably large number of brilliant symphonies, most of which will never be composed; and still a larger number of banal novels, poems, and symphonies, most of which will also (fortunately) never be written. In the Library of Mendel, there are an uncountably large number of genomes specifying potentially viable organisms, most of which will never appear on earth. Each potential masterpiece, each potential work of pure kitsch, and each potentially viable genome is, like *Moby Dick*, surrounded by a sea of near misses. By far the greatest number of volumes in the library consist of pure nonsense.

This metaphor serves to render the problem of how a particular new organism could evolve from a particular existing organism to one of how random changes in the existing genome could produce a genotype that is also viable, and how a series of such random changes could lead to a particular new genotype. If any link in the chain does not specify a viable organism, then the sequence of mutations is impossible. For any new organism to evolve, there must be a path that leads from an existing organism through a series of survivable organisms.

The environmental influences on an organism's development can also be specified, and the environment(s) in which a particular genotype can produce a viable organism can be incorporated into the description of that organism in the Library of Mendel. The logic of the metaphor does not change. For example, several species of organisms can survive in extreme environments such as an atmosphere dominated by methane. For such an organism + environment to evolve from any existing organism + environment, it was necessary that a series of changes in the genomes specifying both the subject organism and the other organisms involved in maintaining the pertinent environment occurred, and, at each stage of the sequence, each of the organisms specified by these genomes had to be capable of surviving and reproducing in the environment they collectively produced and maintained. Otherwise, "you can't get there from here."

Although it is not central to the story I wish to tell here, it is worth noting the consequence for Dennett's enterprise of incorporating environment into the digital specification of an organism. The survivability of the organisms specified by human DNA depends, as Dennett shows, on a cultural as well as a physical environment. (Human infants cannot survive to sexual maturity without a nurturing and protective cultural environment.) Moreover, the organization of any individual's brain cells and their states at any given moment can be specified by a subset of the information in the Library of Babel—a subset that includes the individual's DNA as well as the accumulated physical and cultural experiences of a lifetime. Through this line of reasoning, Dennett proposed to show that a potentially meaningful cultural product can actually occur only if there is a viable bridge from what currently exists (in DNA plus physical environment plus cultural environment) to the potential product. Many of the potential masterpieces will never be produced because the person who might have produced them will take or has already taken a different path. Poet Gary Snyder's sensibilities, encoded in the cells of his physical body as well as in his written works and other artifacts, are unalterably affected by his early study of Native American lore, by his studies in Zen, and by his life in a remote mountain community. The poetry he might have written had he chosen to study Catholic or Sufi mysticism rather than Zen can no longer be written. What we have is very valuable; what we might have had can never be known: You can't get there from here.

Explicit in Dennett's entire argument is a fairly sophisticated rendering of both culture and biology in terms of information theory. DNA is a code; genes are digits in the code—the equivalent of the English alphabet. The cell—starting with the ovum—is a genetic *reader*. Like the decoder in a transmission circuit, the reader must be precisely matched to the code it is to read: DNA and the cell that reads or decodes DNA into protein molecules must evolve together. Dennett refers to evolution as an algorithm—a substrate-neutral, mindless, or mechanistic series of steps that will lead in a deterministic (but not necessarily predictable) way to a result. The idea of substrate neutrality (e.g., the series of steps for computing the square root of a number—the result is the same whether the steps are carried out on paper, via a mechanical computer, or by an electronic computer) is linked to the distinction between message and code, form and content, message and meaning. It is also linked to the traditional differentiation between genotype and phenotype, to which I return shortly.

Following up on a suggestion from Dawkins (1989), Dennett proposes that culture is specified by elemental ideas, which he (and Dawkins) calls *memes*. Memes play a role in cultural transmission similar to that played by genes in biological transmission. For Dennett's argument, this close parallel closes the circle through which the fundamental idea of evolution is applied to culture as well as biology:

> Not only all your children and your children's children, but all your brainchildren and your brainchildren's brainchildren must grow from the common

stock of Design elements, genes and memes, that have so far been accumulated and conserved . . . all the achievements of human culture . . . are themselves artifacts . . . of the same fundamental process that developed the bacteria. . . . (Dennett, 1995, p. 144)

At a detail level, Dennett's argument is metaphorical; even the word *meme*
was chosen to underscore the metaphorical relationship to genes. Yet the
process through which these basic units of culture are transmitted and
shaped into new cultural artifacts, Dennett claims, is literally the same
process as that which governs the transmission of genetic units and their
shaping into new organisms. Evolution is an algorithm and works equally
well in any substrate.

DNA is often referred to as a *genetic code*, and indeed it has several attributes of a digital code. Genes (words) are formed by various combinations of just four nucleic acids (letters). Although this may not be quite as
simple as the "dah-dit" of Morse Code, it seems remarkably similar. The
similarity goes further: Taken in various combinations, genes specify or
contribute to construction of unique proteins. Traditionally, a particular set
of genes, referred to as a *genotype*, is distinguished from the organism that
is made up of the proteins they encode, the phenotype. It is to this duality
that Oyama (2000) primarily objects, as I discuss in the next section. However, it is not evident that Dennett's account hinges on a rigid separation
between genetypic information and phenotypic expression. Indeed he more
or less explicitly incorporates the information in the immediate environment of the cell (the genetic reader) as well as in the more general environment of the organism into the algorithmic process that shapes the intergenerational transmission of genetic information.

To the extent that DNA is regarded as a code and the genome of a particular species as the equivalent of a restricted code (as English is a restricted
form of the general alphabetic code and William Shakespeare's style is an
even more restricted form of English), the fundamental laws of information
theory should apply. The formula for H can certainly specify the maximum
information capacity of DNA in general or of any genus' or species' genome
in specific. As with the specification of maximum information capacity of
human language or of a particular experimental design, whether H is a particularly useful number in this instance is another question.

An obvious application of information theory is suggested by the concept of an evolutionarily stable strategy (Dawkins, 1976, 1998; Dennett,
1995). The theory of evolution poses two interesting problems for the transmission of genetic information. First, drawing on Dawkins' (1976) metaphor
of the *selfish gene* and Dennett's (1995) metaphor of an *evolutionarily stable
strategy*, it is evident that one of the design problems to be solved by species is the problem of exact replication. A complex of genes that includes

coding for accurate, error-free replication, once it had evolved, would re-main stable forever. Moreover, were error-free replication possible, we would expect that it would evolve, eventually, in virtually every evolution-ary line. Thus, even if the early adopters of error-free replication were over-whelmed by more vigorous organisms, eventually organisms would evolve that combined a high level of reproductive success within their own ecolog-ical niche with error-free replication. The result would be an end to evolu-tion—probably long before multicelled organisms appeared.

The fact that evolution has not ceased requires explanation. Shannon's first theorem provides the required explanation. The first theorem tells us that perfect transmission of information—including genetic information—is impossible. There is no danger that any organism will evolve a way to repli-cate its own genetic information perfectly generation after generation. It cannot be done. The first theorem also says, however, that clever use of re-dundancy allows us to come arbitrarily close to perfect, error-free transmis-sion. Through redundancy (including mechanisms for detecting and cor-recting transmission errors) it is possible to maintain a high level of genetic stability over a long period of time—as highly successful species, including several species of sharks, have done. The first theorem also tells us where to look for explanations of long-term genetic stability: Look for various forms of encoding redundancy, including mechanisms for detecting and correcting errors as well as mechanisms for preventing errors from occur-ring in the first place. An obvious form of encoding redundancy is the envi-ronment, which acts as a metaphorical editor, deleting alterations in the genotype that lead to unsuitable alterations in the phenotype, and thereby implicitly reinforcing the stability of a successful design.

But there is also a high degree of variation among organisms in the ac-curacy with which they transmit their genetic information. Some organ-isms are relatively stable over long periods of time; other organisms (unfor-tunately including the viruses responsible for influenza and AIDS) evolve rapidly. If it is possible to develop transmission with such a low rate of errors that organisms such as sharks can swim about, effectively un-changed, for millions of years, why haven't other organisms adopted these techniques? The tradeoffs specified by Shannon's second theorem help explain the energy costs involved in pursuing long-term evolutionary stability. The second theorem tells us that transmission accuracy (and, by implication, long-term genetic stability) comes at a cost. Accurate trans-mission of genetic information requires a higher channel capacity (more DNA, more energy devoted to preventing and correcting genetic errors), a lower transmission rate (production of fewer offspring), or both. The sec-ond theorem also tells us to consider the accuracy/capacity/rate tradeoffs when analyzing the evolutionary strategies of either unusually volatile or unusually stable species.

There seems to be a contradiction here. On the one hand are species like the shark and cheetah, which exhibit long-term genetic stability. On the other hand are species like the viruses responsible for AIDS and influenza, which thrive by virtue of their short-term genetic instability—their ability to evolve new versions faster than the body's immune system can keep up. Therein lies at least part of the answer: At the environmental level relevant to the shark, the ocean changes very slowly. The shark has achieved a high level of perfection as a killing machine, with few natural enemies; any deviation from its basic design would render it less fit. At the environmental level relevant to a virus, the immune system of a human being (or other animal) changes extremely rapidly: The ability to alter relevant features of the phenotype in a handful of generations, conferred by a highly unstable genotype, is essential to these species' long-term reproductive success. To express this concept in terms of information theory is awkward, if not impossible; it is difficult to see what, if anything, would be added to our understanding of these phenomena by formal invocation of information theory.

The information theoretic account poses several other problems, which I discuss in the next section. In brief: Once again, the implicit transmission metaphor needs to be justified. In effect, the transmission metaphor implies that a set of chromosomes is a message encoded in DNA (by the DNA in the parent cells or perhaps by the selective pressures of the environment acting on the DNA) and decoded by the cell. So closely does DNA resemble a code transmission system, that it is easy to miss that it is not necessarily so. On the other hand, if "DNA AS TRANSMISSION CIRCUIT" is indeed metaphorical, that does not necessarily imply that it is not a useful metaphor, that its metaphorical entailments are inexact. It does warn us to inspect each of these entailments closely, one by one, before accepting them.

THE CELL AS A DECODER

Oyama (2000) criticizes the computational model vigorously, primarily on the grounds that it smuggles dualism back into evolution theory and renders both the cell and surrounding environment as passive recipients of DNA's shaping action. I have focused on Dennett's popularized exposition, with its metaphor of the cell as a "genetic reader," because the problem is most obvious there and because Dennett relies so heavily on the computational model in extending evolution theory to human culture. Oyama's objection seems to be based on the idea that the entire chemical environment within the cell, and indeed the external environment with which the cell interacts, are all part of the unit of evolution. The "genetic reader" or "genetic decoder" is not a passive Turing Machine, reading the tape of DNA and responding by cranking out whatever proteins are called for. To the contrary, the cell is a complex mixture of chemicals, many of them considerably more reactive than DNA. The only way, Oyama claims, to overcome dualism is to conceive of the process as an ongoing and purely chemical interaction in which DNA plays but one part, often a minor part. As noted in the preceding contrast between sharks and viruses, both stability of the phenotype across generations and many of the variations within a phenotype can be explained by the interaction of cell chemistry with environmental conditions, with no reference to coded messages.

It is helpful to think again about the problem of perfect transmission. For the evolutionarily stable strategy of perfect genetic transmission to have meaning, the environment in which the organism lives and reproduces would also have to be stable. Otherwise as the conditions in which it seeks food and evades enemies change, the organism would become gradually more unfit. From this musing, it becomes apparent that some of the information affecting evolution is in the environment. It is true that perfect genetic transmission is impossible, but that would only become relevant if the environment were perfectly stable over a sufficiently long period. Again the AIDS virus is a perfect example: Part of the reason for its rapid evolution is that the medical profession keeps changing its environment by introducing powerful new drugs and other forms of treatment.

The concept of information is still relevant at some level if only because DNA is so obviously digital. But as with the information capacity of human speech, it may not be useful to have a measure of the information capacity of DNA if other constraints keep the actual transmission rates well below the maxima computed from Shannon's theorems. There has recently been an amusingly dismayed reaction to the news that the human genome may consist of as few as 30,000 to 60,000 genes. However, if the genes interact with each other in pairs, the potential information capacity of even as few as 30,000 genes has an upward limit of 450 million bits. If the genes interact in larger groups, and if the phenotype is determined by a complex interaction of environment with cell chemistry, including DNA, the upward limit is much, much higher. The information capacity of the genome, as measured by H, may simply not be a useful number, and the constraints suggested by Shannon's (1949) theorems may not be the most relevant constraints. Moreover, Oyama's (2000) critique—by dispensing with the separation of DNA from surrounding cell chemistry (as well as from the environment generally) and dispensing with the separation of information from computation—negates the linear model on which information theory, as we ordinarily think of it, is based. There is no source, no receiver, no encoder, no decoder. All that is left is something with some of the surface characteristics of an encoded message. But even this much of the metaphor fails because a genome expressed in DNA does not seem to act like a message expressed in a medium in any conventional sense.

Dennett's (1995) exposition, in which DNA is like a computer tape and the cell is a mechanistic reader or transcriber of the tape, implies that DNA determines the entire outcome of the process, with the cell relegated to the role of a transcription machine (decoder) and the environment relegated to the role of selective constraint. Alternatively, one can think of the cell or environment as the active principle, with DNA as constraint. But as Oyama (2000) points out, the chemistry of the cell is complex and unitary; DNA is embedded among a complex variety of reactive chemicals. Thus, if Oyama's account is correct, information is not merely in the DNA; it is also in the chemistry of the cell and the surrounding environment. The organism is a product of the entire system, and it is misleading to separate DNA from the rest of it and treat DNA as the *message* and the rest of the cell as the *receiver*.

Oyama seems to reject the traditional phenotype/genotype distinction; it is not clear how seriously this is to be taken. Recent results in cloning experiments certainly undermine a simplistic application of the *code/reader* metaphor. As experience with cloned animals builds up, it has become clear that, despite strong surface resemblances between parent and offspring, the replication is rarely exact. Cloned offspring rarely survive, and the survivors often exhibit deficiencies that undermine their immediate health, their ability to reproduce themselves, or both. Sometimes these deficiencies do not show up until well into adulthood. Humphreys et al. (2001) hypothesize that cloning bypasses the normal way in which chromosomes from the two parents interact, and thereby causes some genes to be expressed at abnormal levels (see also Best & Kellner, chap. 8, this volume). Not only do genes interact in a complex way with the chemistry of the cell, but DNA from the two parents interacts in subtle and as yet poorly understood ways. Moreover, the chemistry of the cell, the DNA "reader," is itself specified by DNA. All this is a far remove from the long-cherished metaphor of a *genetic code* and the resultant idea of cloning as a process much like printing endless exact copies of an organism with a sort of cellular mimeograph machine. Indeed the concept of a *gene*, which originates in Mendel's pre-DNA breeding experiments, is best regarded as a simplifying metaphor for what is coming to be recognized as a complex set of chemical interactions (for detailed discussion see Braman, chap. 4, this volume).

Still evolution does happen, and DNA plays a key role. DNA interacts with the environment, both within the cell and externally, to maintain relative stability of a species from generation to generation. Random changes in DNA interact with changes in the environment to bring about adaptive modifications in the species, eventually, in more extreme circumstances, resulting in the emergence of an entirely new species. *Transmission over a circuit with noise* is a good metaphor for this process and, at some level, is a literal description. To that extent, at least, the theorems of information theory must apply if only to describe the boundaries within which the system must operate.

IMPLICATIONS: INFORMATION, EVOLUTION, AND DUALISM

In electrical engineering, Shannon's theorems have been marvelously productive, allowing engineers to pack a truly incredible amount of signal into a single circuit—as attested to by the high-speed DSL modem that shares a single wire with my telephone. By the same token, Shannon's H and the associated theorems can be used to estimate the maximum capacity of language, of neurons in the human brain, or of any species' genome. The resulting numbers are huge beyond comprehension—a fact that has been celebrated by any number of new-age mystics who use it as a basis for proclaiming the possibility of our every fantasy. However, the real lesson is rather more sober.

Although as discussed earlier the plot of many a novel or movie turns on an incomplete or garbled transmission, it does not appear that the constraints imposed by channel capacity, transmission rates, and the tradeoff between accuracy and transmission rates play any important role in either the successes or failures of everyday human communication. All human communication is, in principle, subject to Shannon's (1949) theorems, and the energy that fuels the activity of our neurons during communication is in principle subject to the laws of thermodynamics. However, these constraints tell us nothing about how people actually communicate or why they so frequently fail to do so successfully. The theorems associated with information theory have been of some metaphoric use, but the metaphors based on information theory are probably misleading as often as they are insightful.

If I understand Oyama's (2000) critique correctly, it appears that a similar point holds with respect to evolution. Given the amount of junk DNA on a typical chromosome, it does not appear that transmission capacity per se is an important constraint on either intergenerational stability or intergenerational change. Junk DNA consists of long sequences of nucleic acids that serve no discernible biological function. Because it also does no harm, it is not selected against and so tends to accumulate over many generations. If transmission capacity were a constraint on the reproductive fitness of a species, then by definition junk DNA would be harmful (because it would soak up valuable and needed transmission capacity) and would tend to be selected against, hence it would not build up to such a degree. Genetic stability and variety are governed not by the capacity limits of a genetic code, but by the chemistry of the cell and the rigors of the external environment. Dennett's (1995) metaphor of the Library of Babel, in which the vast majority of volumes are nonsense, tries to make much the same point. But as it does so, it reintroduces the form/matter dualism to which Oyama so vehemently objects.

Information has a literal meaning with respect to human communication that predates and seems to legitimate Shannon's (1949) appropriation of the term. At first glance it appears that Shannon's H simply quantifies this familiar concept, as a number of scholars have argued, beginning with Weaver (1949) and Schramm (1955). However, for Shannon's H to work as a literal quantification of the everyday concept of information, it would be necessary to specify a well-defined distribution of alternatives to a given fact, datum, or lore—an objective that proves elusive in practice (Ritchie, 1991). We are left with the conclusion that Shannon's use of the term *information* is a metaphor based on the everyday understanding, a metaphor ironically prefigured by Eliot (1934), "Where is the wisdom we have lost in knowledge? / Where is the knowledge we have lost in information?" Communication scholars eager to capitalize on the success of communications engineers in the heady postwar environment of the 1950s turned this metaphor back on itself, making the mathematical concept of information a metaphor for human information in much the same way that the statistical concept of probability has been made a metaphor for epistemological probability (Ritchie, 2003).

As I show elsewhere (Ritchie, 2003), *probability* began as an epistemological term, distinguishing mere opinion, which was based on reasoning from observed evidence, from true knowledge, which was based on reasoning from first causes or from authoritative sources, such as Aristotle or Holy Scripture. *Probability* referred to the strength of an opinion based on the strength of supporting evidence. As the science of statistics developed, statisticians linked the distribution of expected outcomes to the idea of probability, at least in part in an attempt to develop a way to quantify arguments in legal cases. Over time the concept of *probability* came to be understood in terms of the distribution of expected outcomes in a chance-governed process. Scientists, especially social scientists, who deal almost exclusively with processes that have indeterminate outcomes, then began to use probability as a description of the strength of an experimental finding. This epistemological use of probability has been generalized in everyday uses as a metaphor for our expectations of future events even when we have no basis for statistical computations, as for example when we say "the odds are about ten to one that I will miss the party" or, almost interchangeably, "I will *probably* miss the party." Thus, the concept has come full circle.

Like the *probability* metaphor, there seems to be nothing intrinsically wrong with the *information* metaphor as long as we recognize that we really have no way to apply mathematical information to the task of measuring epistemological information. Yet the two concepts are easy to conflate—and indeed it is difficult to avoid conflating them—precisely because of an underlying literal relationship.

In biology the concept of information can be traced to the Aristotelian theory that male sperm *informs* the neutral medium of the female ovum, and I believe it is this conflation to which Oyama (2000) most strongly objects. The classical view of heredity explicitly renders the female as wholly passive and the male as wholly active. Modern biology has long since moved beyond this simplistic and profoundly sexist view of heredity, recognizing that genetic material is contained, in more or less equal measure, in both egg and sperm. However, the model of DNA as message and cell as receiver/transcriber continues the same basic model in which an active informative principle is separated from a passive medium and ignores the contribution of the information contained in the chemistry of the ovum. The basis for Oyama's whole-hearted rejection of the "DNA AS INFORMATION" metaphor is that the cell is a highly reactive and complex chemical environment, and DNA is only a small part of this environment. If the role of DNA is conceptualized in terms of information, we also need to recognize the information embodied in the overall chemistry of the cell, not to mention in the surrounding environment. The "GENETICS AS INFORMATION THEORY" metaphor seems to work against this recognition because of its connection with the source–message–receiver model.

As is the case with human communication, conflation of metaphoric with literal understanding of genetic processes is encouraged by the underlying literal connections with information theory. DNA is in a real sense a code, and a strand of DNA looks remarkably like a computer data tape. Thus, it is all too easy to make the leap from understanding part of the genetic process as information in a literal sense to understanding the entire process as an instantiation of information theory.

I have previously argued, based on Vervaeke and Kennedy (1996), that it is reasonable and even useful to draw on multiple root metaphors in understanding a complex concept. I have also argued that two major tasks of metaphor analysis are to tease apart the metaphoric from the literal, and to sort out the useful and applicable metaphorical entailments from the misleading and inapplicable entailments. Information theory does apply literally to the genome in some sense at some level. However, if genetic processes are controlled by constraints other than the theoretical maximum transmission capacity of DNA, its literal application may not be particularly useful. Contra Oyama (2000), I also believe that information theory can be useful as a metaphor to help explain the complex processes of heredity. As

Oyama insists, the use of information theory as a metaphor for genetics is tricky and, if poorly handled, must inevitably lead back to the kind of active/passive, form/matter dualisms associated with the Aristotelian theory of heredity. Whether Oyama is correct in her claim that the linkage between information theory and dualism is so strongly habitual in our culture that the *information* metaphor cannot be used without opening the door to dualism remains to be seen: She certainly makes a convincing case for it.

This leads us back to the question of *information theory* as a metaphor for human communication. As I commented in the preceding, there is really nothing wrong with using mathematical information as a metaphor for epistemological information as long as we recognize that we have no way to measure the actual mathematical information content of an epistemologically informative message. However, extending Oyama's complaint about the information metaphor in genetics, the application of mathematical information theory as a metaphor for the expression of meaning lends itself all too readily to a dualistic separation of meaning from both messages and communicative contexts. Given the history of theorizing about human communication over the past 40 years, one would be justified in extending Oyama's (2000) pessimism about the prospects for separating information metaphor from dualism in biology to the prospects for separating the information metaphor from dualism in human communication.

Given the growing evidence that genetic material does not have a simple one-to-one mapping with expressed features of the organism, and that the concept of a gene is probably best viewed as metaphorical, at the very least our understanding of these processes clearly requires some refinement. I have a high degree of admiration for Dennett's (1995) ambitious attempt to develop a theory of cultural evolution as an outgrowth of the theory of biological evolution, but it seems increasingly clear that he (and others with similar interests) run the risk of getting caught up in their own metaphors. Dennett has elsewhere made it clear that he does not believe in the kind of mind/body or form/matter dualism to which Oyama objects so strenuously, yet his use and expansion of the information metaphor seems to implicate his ideas in precisely that kind of dualism.

It would seem to be a responsibility of rhetors who use a complex metaphor such as information, as well as of analysts who interpret and evaluate such usage, to do so with careful attention to the entailments of these metaphors. This process must begin with carefully teasing apart the metaphorical from the literal. Once it is clearly understood how a use of information theory is metaphorical, it is necessary to consider carefully each of the entailments: Which of the related metaphors is apt, which is inapt, and which is downright misleading? If one is to draw on and even extend the information metaphor, as Dennett (1995), Dawkins (1976, 1998), and many other writers have done, to render a complex concept more comprehensible,

there is a considerable responsibility to delimit the entailments and specify
the various ways in which the metaphor may mislead. If one is to extend
the information metaphor as a basis for developing the evolution process
as a metaphor for cultural processes, as Dennett (1995) and Dawkins (1976,
1993, 1998) have both done, it is even more important to be clear as to when
the discussion is intended to be taken literally, when it is intended to be
taken metaphorically, and, again, where the metaphor may be inappropriate or even misleading.

3

Conditional Expectations Communication and the Impact of Biotechnology[1]

Steven S. Wildman
Michigan State University

Biotechnology is a potent source of hope and fear, both of which are illustrated in this book. At the core of each is the recognition that we are witnessing the creation of a technology that in the most literal sense is powerfully transformative. It is capable of transforming significant features of the living world around us and, because we are biological beings, capable of transforming its creators as well. Both the long-term environmental implications and the ethics of biotechnological interventions to alter the traits of individuals of other species and members of our own are hotly debated. The existence of factual support for arguments made on both sides of the debate makes it particularly difficult to resolve. Biotechnology advocates point to the benefits of the higher yields of genetically modified crops and the human suffering that can be alleviated by correcting medical problems caused by genetic defects. Biotechnology's critics counter by pointing to the dangers of unintended consequences in the larger ecological and societal contexts in which modified organisms are introduced; and the thought that parents might employ technological measures to determine (or at least influence) traits such as height, athletic ability, intelligence, or physical attractiveness, for their unborn offspring—the so-called *designer baby*—is deeply troubling to many.[2]

Given such concerns, we might ask whether communication is a human trait that can be altered with biotechnology. There are at least two ways in which this could happen. Biotechnology might be used to alter the genetic foundations of communication, or biotechnology might be employed as an instrument of communication. It seems appropriate that the question of biotechnology's potential impact on human communication be addressed in a book on biotechnology and communication. Of course there is a trivial sense in which this question must be answered in the affirmative. Human communication is dependent on a variety of traits and abilities, such as intelligence and quality of voice, that are at least partly genetically determined. Therefore, intervention to alter these traits and abilities would necessarily affect the way an individual communicates. This answer is trivial in part because it is obvious, but, more important, because it depends in no way on an understanding of what it means to communicate. A more fundamental answer to the question of whether genetic intervention might change the nature of human communication requires a definition of communication sufficiently rigorous to support both logical and empirical analysis.

The next section argues for a conditional expectations definition of communication that is justified by the assumption that behavioral agents select among behavioral options on the basis of beliefs or expectations that are conditional on information they perceive regarding the likely impact of alternative behaviors on their well-being. This definition will underpin the arguments and analyses developed in the remainder of the chapter, including an analysis of the role of natural selection in shaping communication that forms the basis for my response to the question of whether biotechnology might modify human communication at the end of the chapter.

Although the formal definition of *conditional expectations communication* requires some setup, the basic intuition underlying the definition and the analytical framework built on it is straightforward. Sense organs and an ability to interpret sensory information have obvious advantages in a world governed by competition for scarce resources. It may not be appropriate to attribute sensory capacities to plants analogous to those in animals, yet it is now widely accepted that plants process and respond to information of various types about their environments, including information from other plants. For example, studies of wild tobacco plants have shown that they respond to chemicals released by injured plants of other species nearby by increasing their own production of chemicals noxious to insects that feed on them.[3] Thus, it is fair to say that most, if not all, living beings process information about their environments to aid them in selecting strategies that better promote their interests. Strategies selected in this manner are conditional on the information received and must, at least implicitly, reflect an expectation regarding the state of the environment in which the strategies will be employed. Communication occurs when the information processed is supplied by another agent in the environment motivated to do so by some interest that it is pursuing with respect to the agent processing the information.[4] That is, the information is supplied to elicit a response by another agent that is beneficial to the first.

The conditional expectations definition of communication has a number of analytical implications that should be of interest to communication scholars generally. One is that an analysis of a communication practice cannot be complete unless the interests of all parties involved are considered. If the practice is one that persists over time, then both the provision of the information and the responses to it must be seen as the best available strategies, respectively, by the sender and receiver of the information. Conversely, when a heretofore stable communication practice is abandoned, it must be the case that either the provision of the information or the response to it is no longer viewed as a best strategy by the sender or receiver, respectively.

A second implication is that at some level each participant in a stable communication process selects its strategy based on an understanding, whether explicit or implicit, of the conditional expectations logic driving the other participants' selection of their strategies. For communication practices that are the product of learning by individuals, this knowledge is developed as the communicating agents interact with each other. So in this sense this knowledge is socially constructed. (Note that this claim is not specific to human communication.) However, as the work on animal signaling reviewed below shows, there are many communication practices that are instinctual in nature. In such cases, the knowledge resides in the genome of the species (or genomes for communication between members of different species). But even for communication practices that are learned, it is fair to say that the ability to form conditional expectations (including expectations of how others will respond to information a communicating agent supplies) and select strategies based on them is a product of genetic selection.

A third implication follows from the generality of the description of conditional expectations communication (CEC). The range of interests pursued through communication would appear to be limited only by the range of interests of the agents involved. They could be as diverse as physical survival, procreation, profit, social status, and winning a soccer game. Thus, the CEC framework can be used to analyze all sorts of human and nonhuman communication practices; when the contexts in which communication practices arise are analytically similar, we can expect to find similar communication strategies employed by both human and nonhuman actors—the communication equivalent of convergent evolution. The highly similar, costly signaling models reviewed later in this chapter, which were developed independently by economists and evolutionary biologists to explain communication practices in very different settings, illustrate this point.

A fourth implication is that any process by which the expectations governing responses to environmental stimuli are formed that is consistent with selection for fitness may constitute the basis for the development of a communication practice. Genetic selection for instinctual behaviors, learning based on individual experience, and the social inculcation of beliefs would be included among such processes.

A fifth implication is that, because communication practices based on conditional expectations should have adaptive advantages in any selectionist system in which the experiences of individual agents modify behavior (either through agents modifying their own behaviors or through the selection of agents with different behaviors), we can think of the logic of CEC as a force shaping the direction of evolutionary change. Just as selection for fitness will favor individuals and species better adapted for dealing with the force of gravity, so it also will favor individuals and species that better exploit the logic of conditional expectations in their communication practices.

A CONDITIONAL EXPECTATIONS MODEL OF COMMUNICATION

All living organisms respond to variations in features of their environments that affect their well-being. The ability to do so is certainly an aid, if not a prerequisite, to survival. In an evolving biological system, selection for fitness will produce responses that, over time, do a better and better job of promoting the biological success of the individual organism[5]—at least for variations that are of a recurring nature. Because a response necessarily lags somewhat behind the stimulus that elicits it, adaptive responses must actually address conditions the organism expects to obtain at the time of the response. Thus, there is a predictive character to stable stimulus response systems. The variations in the environmental features to which organisms respond constitute the information on which the predictions are based. Although the processes through which responses evolve may or may not be conscious, the end result is organisms that select responses to information encountered in their environments on the basis of: (a) (subjective) probabilities (or expectations) assigned to the various outcomes that could attend each of the alternative responses that might be employed,[6] and (b) some sense of the contributions different outcomes make to goals these organisms pursue[7] (hereafter referred to as importance weights, or just weights, assigned to outcomes). The process of response selection conditional on information encountered in the environment is illustrated in Fig. 3.1.

As an example of conditional response selection, consider a deer hiding in a thicket at the edge of a meadow. The most nutritious forage is in the meadow, but the danger of being attacked by predators is greatest there also. The deer's options are to stay in the thicket where the odds of being attacked are always lowest, but remaining hungry is a certainty, or to ven-

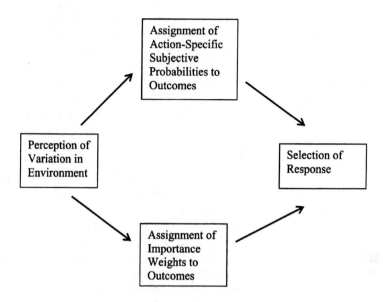

FIG. 3.1. Conditional expectations response selection.

ture into the meadow where the odds of falling victim to a predator are higher, but where the forage needed to maintain its vitality in the long run is available. Detection of the scent of a predator would alter the probabilities the deer assigns to the possibilities of falling victim to attack or grazing peacefully in the meadow should it decide to leave the thicket. Observation of the state of the foliage in the meadow, which would enable the deer to better gauge the nutritional benefits of grazing there, may also influence the deer's choice, as might the amount of time elapsed since its last good meal. Thus, the weights (goal contributions) assigned to outcomes, and their influence on behavioral choices, may vary independently of probabilities assigned to outcomes for different behaviors.

For most organisms, other organisms with which they interact are critically important elements of their environments. This is obvious in the deer–predator example. The deer is a prospective meal to the predator, and the predator is a threat to the deer's life. Neither party is indifferent to the outcome of their interaction. When organisms are not indifferent to the character of their interactions, natural selection favors those organisms with the most effective strategies for influencing the outcomes of their interactions with other organisms in ways that promote their own interests, which could include strategies that influence the behaviors of the organisms with which they interact.

Providing information to which another organism might respond could be one strategy for influencing its behavior, and it seems natural to classify such a strategy as communication. If behaviors are selected on the basis of conditional expectations and communication is the provision of information to elicit behaviors preferred by the suppliers of information, then communication practices that are stable and persist over time must be consistent with the process (or processes) by which conditional expectations are formed. This suggests that communication might fruitfully be studied from a conditional expectations perspective, and that a conditional expectations definition of communication would facilitate such an exercise. Such a definition is offered shortly, but it is helpful to first refine the notion of what constitutes information within this analytical framework.

No organism is capable of responding to variation in every feature of its environment. In fact most features of an organism's environment fall outside its perceptual thresholds. Sense organs and the ability to perceive features of the environment evolved because they enabled organisms to respond to variations in elements of their environments that affected their well-being in ways that improved the odds that their genes would be represented in future generations of their species. Of course the environmental elements for which an ability to respond is important vary considerably among different types of organisms. Although it may be important to an antelope to know whether the pride of lions on the horizon is actively searching for prey or just out for a stroll, this is not likely to be a matter of concern to a nearby elephant or a mouse underfoot because neither are vulnerable to attacks by lions. One would thus expect antelopes to be sensitive to behavioral clues to a lion's intentions to which elephants and mice are totally oblivious. The distinction between the features of its environment for which an organism can and cannot perceive variation motivates the following definition of *informative*.

> A feature of its environment is informative for an agent if it varies in a way that is perceivable by the agent and the agent's knowledge of the state of the feature influences the agent's assessment of the outcomes likely to attend alternative courses of action.

For the remainder of this chapter, I use the term *communicative* to refer to an act by one agent that modifies an environmental feature informative to a second agent if the intent in doing so is to elicit a response beneficial to the first agent. I say that the first agent is communicating with the second if the second agent is aware of the communicative act. This can be referred to as the CEC model. More formally:

> A communicative act is the nonaccidental creation or modification by one agent of one or more features of the environment that are informative to a second agent (or collection of agents) in a manner that, if perceived, influences the second agent's assessment of the likely consequences of alternative

courses of action in such a way as to increase the likelihood that the second agent will pursue a course of action that favors some interest of the first agent.

Communication occurs when the environmental feature modified by a communicative act is perceived by the agent it is intended to influence.[8]

The requirement that an act generating a response beneficial to the actor be nonaccidental to be classified as communicative reflects an assumption that an agent attempting to communicate with one or more other agents is acting on the basis of conditional expectations. If predictions underlie behaviors selected on the basis of conditional expectations, a communication practice that effectively exploits the process of conditional expectations response selection must rely on a prediction—a prediction of likely responses to the information supplied. Furthermore, given that the agent supplying the information might pursue other courses of action, it must see the choice of a communicative act (and the act chosen) as its best response to the information it has about its environment. Therefore, if a particular communication practice is to be stable over time for the agent supplying the information, the nature of the response must be such that supplying the information continues to be a preferred strategy while the response that warrants supplying the information is also a preferred strategy for the agent responding. That is, for both agents, the outcomes experienced must not depart so dramatically from those predicted (either in terms of the likelihood of their occurrence or their importance to the agent experiencing them) as to lead to revisions of the outcome probabilities and/or the weights assigned to outcomes large enough to warrant selection of a different response to the same information in the future.

The role of realized outcomes in the assignment of information-conditional subjective probabilities to the possible outcomes associated with different behaviors and the assignment of weights to outcomes is illustrated in Fig. 3.2. Figure 3.2 can be used to describe the situation of either the initiator or an intended respondent to a communicative act. For an agent trying to communicate with a second agent, the communicative act is its response to information it perceives as suggesting that the odds of a favorable response to the communicative act are high enough to justify the attempt at communication. The actual outcome is the second agent's behavior following the communicative act. If Fig. 3.2 is used to describe the situation of the second agent, then the first agent's communicative act produces the variation perceived in a feature of the environment to which the second agent responds (if the variation is noticed). The outcome associated with the response could relate to the consequences of a subsequent interaction with the first agent, although this is not the only possibility.[9] The full set of relationships between the two agents is depicted in Fig. 3.3.

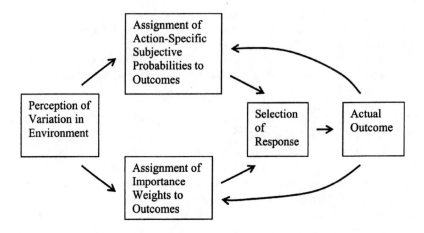

FIG. 3.2. Influence of actual outcomes on assigned probabilities and outcome weights.

Different types of learning processes might be represented by the arrows connecting actual outcomes to subjective probabilities and outcome weights. Thus, individual agents might adjust their subjective probabilities in response to variation in the frequencies with which different response–outcome combinations are observed, or they could change the weights assigned to outcomes on the basis of outcomes observed or experienced. For example, for an individual deer, an increase in the local deer population should reduce the likelihood that any individual deer will be attacked by predators while overgrazing might reduce the quantity and quality of grass in the meadow. Over time experience in this altered environment should lead to a downward revision in the probability assigned to being attacked in the meadow and a simultaneous lowering of the nutritional valued assigned to grazing there. How this would affect the likelihood of entering the meadow when the scent of a predator is in the air would depend on the relative magnitudes of the two adjustments in the deer's response function.

Modification of subjective probabilities and outcome weights in response to lessons personally learned requires that individual agents have sufficient opportunities to trial different responses to the same information to benefit personally from trial-and-error learning. This is not always the case. In some situations, responses not well adapted to variations encountered in the environment may have fatal consequences, or the relevant variations in environmental features may occur infrequently in the life of a typical individual. In such situations, genetic selection for fitness may play the same role in the development of communication practices governed by conditional expectations as does personal learning when individual agents modify their behavioral strategies in response to personal experience. In

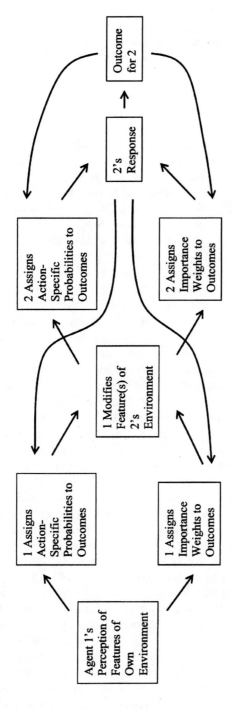

FIG. 3.3. Feedback from experience with conditional expectations communication for two agents.

one case, an understanding of the conditional expectations logic is preserved in the memory of the individual actor; in the other, it is preserved in the genome of the species and reflected in genetically determined behavioral tendencies.

Broader Applications of the CEC Model

The discussion of CEC to this point has been motivated by examples of animals interacting on the basis of behaviors selected for biological fitness. Yet the basic principles illustrated have much broader application. The critical terms in the definitions of *information, communicative,* and *communication* need not be restricted to animals[10] or to strictly biological settings. For example, the communicating agents could be firms and consumers interacting in a product market, and the environmental features to which they respond might be economic factors such as prices and advertisements. Another example would be people generating and responding to status cues in a social context. Individual learning can still form the basis of expectations formation in these settings, but expectations might also emerge through the social processes that give rise to norms, traditions, and generally accepted rules of thumb. (It is worth noting that even if individuals accumulate enough experience to benefit from trial-and-error learning, when the appropriate response to a unit of information is stable across generations, there is an advantage to the individual of having the appropriate response be automatic [biologically or culturally instinctive] because there are personal costs to the poor responses that likely will be trialed before more adaptive responses are discovered.[11]) The basic framework should generalize to any situation in which: (a) agents capable of perceiving and responding to variation in features of their environments interact and pursue goals dependent on those interactions, and (b) tendencies toward fitter responses are preserved—either as individual agents modify their behavior over time or through environmental culling of those with less fit responses. In social and economic contexts, norms, traditions, and institutionalized forms of learning can play the same role as genes in biological systems as vehicles for the transmission of knowledge of fitter responses and communication behaviors from one generation to the next.

As mentioned in the introduction, the generality of the CEC model of communication suggests that similar problems and objectives should be addressed with structurally similar communication practices regardless of the types of agents involved. Thus, we should not be surprised to find a variety of human communication practices mirrored in communication practices employed by other species. I believe this explains the independent development of similar, if not identical models, of costly signaling by economists and evolutionary biologists, as well as recent work on social norms. The findings of this literature are reviewed, and its implications for a more general study of communication assessed, in the next three sections.

Costly Signals and the Problem of Dishonest Claims

Signaling has been offered as an explanation for distinctive features of activities as diverse as mate selection in cockroaches (Moore et al., 1997), predators' pursuit of prey (Zahavi & Zahavi, 1997), wages paid different job applicants (Spence, 1974), and expenditures on advertising by profit-motivated enterprises (Bagwell & Ramey, 1993; Nelson, 1974, 1975). Each of these situations is characterized by two distinct types of agents (male and female cockroaches, predators and prey, employers and job applicants, and buyers and sellers), and in each of these situations an individual member of one of the two types selects one or more members of the other type of agent with which to initiate some sort of interaction. A female cockroach selects a male with which to mate,[12] a predator selects a specific member of a prey species to pursue, an employer selects among job applicants and sets starting wages for new hires, and a buyer selects among the products of competing sellers when making a purchase. In each case, candidate recipients of the selecting agents' attention are not indifferent as to whom is chosen. All male cockroaches want to mate; each member of a prey species would like to see predators chase someone else; job applicants all want to be hired and at the highest wage attainable; and every seller wants prospective buyers to select its products rather than its competitors'. Thus, we should not be surprised if prospective selectees seek ways to persuade selecting agents that their interests are best served by choices that also favor the interests of the selectee attempting to influence them. For a male cockroach, the message to female cockroaches would be, "I am the best available father for your future offspring." For a rabbit, the message to a fox might be, "Other rabbits are easier to catch than me." Similarly, job applicants and advertisers want employers and buyers, respectively, to believe that they are the best among the options being considered.

Note that in each of these situations every prospective selectee will want to make the same best (or worst) option claim regardless of whether it accurately depicts its status. Yet if all prospective selectees represent themselves identically, the claim will be ignored because it cannot be used to distinguish one from another. The models developed by economists and scientists studying evolutionary processes discussed in the next two sections show that signals that can be provided only at some personal cost to the signaler may accurately (or honestly) represent a signaler's status if the level (or amount) of the signal is variable and the cost of the signal is lowest for those signalers for whom the claim of superiority is most accurate. When this condition is satisfied, there will be situations in which the truly superior prospective selectees can benefit on net from signal recipients' responses while setting the signal at a level that their inferior counterparts find too costly to match. The benefit realized by the winning selectee is what gives the signal recipient the power to demand that the signal be provided. At the same time, because the signal accurately sorts selectees, signal recipients' expectations will be confirmed if they believe the superiority claims of the prospective selectees with the highest levels of the signal.

To illustrate the intuitive logic of costly signaling models, consider the problem faced by a hypothetical professional tennis player—a low-ranked traveling professional who unexpectedly loses his regular doubles partner for a charity-sponsored tournament he has already entered. Rather than withdraw and give up any chance for prize money, he puts out the word that he is seeking a partner from among the local teaching professionals. Three—Marty, Tom, and Bill—offer to be his partner for the tournament.

Experience with teaching pros at other tournaments has taught the traveling pro that some may be good enough to give him a decent chance of winning the tournament if they are on their game. Unfortunately, this depends on factors he cannot easily observe, such as their overall level of conditioning and how hard they have been working to develop their skills beyond the level required to effectively teach amateurs. For the local pros, the incentive to play is the chance to share in the prize money of $2,000, which by convention is split 50/50 by the partners on a winning doubles team. To choose among Marty, Tom, and Bill, the traveling pro announces that he will partner with the one who agrees to donate the most to the sponsoring charity, with each allowed to up his promised contribution until all but one drop out.

To see why the amount a local pro offers to contribute to charity might be a credible (or honest) signal of his value as a doubles partner, suppose that Marty has been training hard for the last month and estimates his chances of winning the tournament on a team with the traveling pro at 50%. In contrast, Tom has recently started working a second job, which has cut into his practice time. He estimates his odds of winning at 40%. Bill, who has just returned from a 2-week cruise where partying, not practice, occupied most of his time, estimates his odds at 30%. If we assume that each is willing to pay up to the cash equivalent of his expected earnings in the tournament for the chance to play with the touring pro,[13] then Bill would be willing to give the charity as much as $300 ($300 = 0.7 × $0 + 0.3 × $1,000), Tom would be willing to contribute up to $400 (0.6 × $0 + 0.4 × $1,000), and Marty would be willing to contribute a maximum of $500 (0.5 × $0 + 0.5 × $1,000). As long as each has a chance to respond to the others' offers, Marty, who gives the touring pro the greatest chance of winning the tournament, should bid the most because any donation greater than $400 and less than $500 would exclude the other two and still allow him to benefit on net (in terms of expected value) from partnering with the traveling pro. The sizes of their promised charitable contributions should thus serve as reliable indicators of the teaching pros' relative values as doubles partners for the traveling pro.

Although simple and contrived, this example illustrates two critical properties of signals that may be accepted by signal recipients as credible (or honest) in situations in which there are signalers who might benefit from misrepresenting their value to a selecting agent. The signal is costly to its sender, and the benefit to the receiver of believing the signal (or the message it supports) is greatest for the sender who benefits most from providing it.

This example used monetary measures of the costs and payouts of signaling, but a signal with nonmonetary costs could have been employed even in this situation. For example, if the three teaching pros differed only in stamina and conditioning, the traveling pro might have asked them to run a race of indeterminate length, choosing as his partner the last to stop running. The expectation here would be that the physical discomfort of a long run would increase most rapidly for the more poorly conditioned athletes. Such a behavioral signal would be similar to many that are described in the animal communication literature.[14]

COSTLY SIGNALS IN ECONOMIC AND SOCIAL CONTEXTS

Signaling formally entered the economics lexicon with an article entitled "Job Market Signaling" by A. Michael Spence (1973). Although there were important intellectual precursors (Riley, 2001), Spence's article is generally cited as the inspiration for what has become a vast economics literature on signaling in market settings. The influence of Spence's work on the economics of signaling[15] was recognized with the 2001 Nobel Prize in Economics, which he shared with two other economists who made important contributions to the economics of asymmetric information, which is the study of situations in which the parties to an economic relationship are differentially informed about factors likely to influence the benefits they realize from the relationship.[16] Variations and elaborations on Spence's model have since been employed to explain a great variety of economic phenomena that could not be effectively analyzed with traditional modeling approaches. A good example is an explanation for certain features of advertising strategies introduced by Phillip Nelson in 1974. Like Spence's paper, Nelson's insights continue to inspire new work and commentary. In this section, I describe Spence's model and then compare it to the signaling models of advertising to show that a similar communication logic may be manifest in different ways in different circumstances. I then more briefly describe other applications of this logic to illustrate the range of economic and social activities to which it has been applied. Riley's (2001) recent and comprehensive review of the economic signaling literature is recommended to those who want to delve deeper into this literature.

Spence's Model of Job Market Signaling

Spence considered a hypothetical situation in which firms have positions to fill, but cannot observe directly the productive capabilities of prospective hires, although workers have this knowledge about themselves. Firms do know, however, that some prospective applicants are much more productive than others. Furthermore, competition among firms for workers forces them to pay workers wages equal to the value of their expected contributions to a firm's output. Thus, firms will pay higher wages to more productive workers if they can identify them. However, if it is not possible to distinguish between the two types before they are hired, the offered wage will be the expected productivity of a randomly selected applicant. Suppose, for example, that a high-productivity worker's contribution to a firm's annual output is worth $10,000, that of a low-productivity employee is worth $5,000, and those of high-productivity and low-productivity applicants constitute 40% and 60% of the applicant pool, respectively. Then a firm would offer a wage of $7,000 (.4 × $10K + .6 × $5K) to all workers if unable to distinguish between the two types.

The extra $3,000 that firms would be willing to pay known high-productivity workers is an incentive for both types to represent themselves as highly productive in job interviews. Because talk is cheap, firms would have good reason to discount applicants' self-descriptions. To claim the higher wage, high-productivity workers need a mechanism for demonstrating their worth to employers that will not be duplicated by low-productivity workers. Spence suggested that education might serve this need if the acquisition of educational credentials was more costly for low-productivity workers than for high-productivity workers even if education did not make workers more productive.

To see how education might serve as a credible signal of productivity even when it makes no contribution to productivity, suppose that educational attainment is measured by the number of units of postgraduate training an applicant has acquired in subjects at least nominally related to the job, and that the cost per unit of postgraduate training is $1,000 for a high-productivity worker and $2,000 for a low-productivity worker. Although the reason for this inverse relationship between productivity and the cost of postgraduate education is not important to the structure of the analysis, it is not hard to imagine scenarios for which this makes sense. One would be that an innate aptitude for acquiring skills that make workers more productive in the jobs being filled also makes it easier to learn and pass tests in subjects taught in the specified courses. Low-productivity workers would then have to put in more time than high-productivity workers to pass the required courses—time that might otherwise be devoted to work, recreation, and sleep. The extra $1,000 is the opportunity cost of this additional time.

Given the $1,000 difference in the personal cost of a unit of postgraduate education for high- and low-productivity workers, any level of postgraduate education between two and a half and five units could serve as a reliable signal of high productivity because high-productivity workers could benefit financially from investing in the signal, whereas low-productivity workers could not.[17] Suppose, for example, that employers offered a wage of $10,000 to all workers with at least three units of postgraduate training and $5,000 to all those with less than this amount. Then high-productivity workers would realize a $5,000 return on a $3,000 investment in education, whereas low-productivity workers would not be able to justify their cost of $6,000 on purely financial grounds. All high-productivity workers would thus acquire the three units of postgraduate training required to demand the higher wage, whereas low-productivity workers would all acquire less than three units. (In fact they would acquire none.) Employers' belief that acquisition of three or more units of postgraduate education distinguishes high-productivity workers from low-productivity workers would thus be confirmed. Three units of postgraduate education would thus constitute a stable equilibrium value for a productivity signal in this job market.

Education in Spence's model is essentially a sorting or screening device, as acquisition of the signal accurately categorizes job applicants according to their productivity. It works because the structure of employer beliefs and the wages associated with those beliefs lead workers to reveal something about themselves (the personal cost of acquiring an education) that varies in a predictable way with productivity. Note that it is not necessary that employers understand why a certain amount of postgraduate education is a reliable signal of productivity, only that they recognize the correlation between them. Similarly, the signaling equilibrium described does not require that individual workers know their own productivities, only that they understand the personal costs entailed in acquiring postgraduate training and realize that wages vary with education. The inverse correlation between the personal cost of education and productivity takes care of the rest. That is, it is not necessary that workers think of educational investments as a mechanism for revealing their true productivities to prospective employers. Workers and firms could even falsely (for this example) believe that the higher wages associated with higher levels of education reflect a payoff to skills acquired through training. Spence's actors behave rationally given their beliefs, but the beliefs do not have to reflect a true understanding of the causal mechanisms tying wages to education.

Signaling Models of Advertising

The use of costly investments as signals of quality is not specific to education, as Spence observed. Advertising is one of the most prominent of the
many topics that have been examined from a signaling perspective since
Spence's article. The seminal work is an article by Nelson (1974), which apparently was developed independently of Spence's job market analysis. Nelson analyzed a situation in which sellers differ in the qualities of the products they offer consumers due to innate differences in their abilities to
create quality products, but consumers are not able to distinguish between
high- and low-quality products through prepurchase inspection. Rather, the
qualities of purchased products are discovered as they are consumed.

Nelson's sellers, like Spence's workers, differ in the quality of the product or service they can sell. Like Spence's workers, all sellers find it in their
interest to claim to offer high-quality products because consumers are willing to pay more for higher quality. Therefore, sellers of high-quality products, like Spence's high-productivity workers, would like to find a way to
claim high quality that will not be imitated by sellers of low-quality products,[18] and Nelson's consumers, analogous to Spence's firms, would like to
be able to distinguish between high- and low-quality products before buying them.

Nelson argued that advertising could serve this purpose for goods consumers purchase repeatedly if consumers make repeat purchases only
from sellers whose products have been revealed through experience to be
high quality. In this situation, sellers of high-quality products can advertise
and average the cost of their advertisements over the initial sales generated by their advertising plus the follow-on purchases by satisfied customers. The ad costs per unit sold will thus be higher for sellers of low-quality
products, whose customers will not buy from them more than once. Nelson
showed that when the cost of advertising per unit of product sold varies inversely with product quality, there may be market equilibria with advertising budgets that can be covered out of earnings on sales by sellers of high-
quality products, but not by sellers of low-quality products.[19] Advertising is
a signal in this model and the critical message conveyed to buyers is that
the seller believes it can profit from advertising at the signaling level (Nelson, 1975).

Because Nelson assumed that sellers able to produce high-quality products most easily would do so, advertising provides information about the
value of an innate trait in his analysis, just as educational attainment does
for worker productivity in Spence's model. Subsequent work on signaling
models of advertising has allowed product quality to be a choice variable
for sellers (Bagwell & Ramey, 1993; Rasmussen & Perri, 2001). A seller's
claims for the quality of its product are thus a promise to incur the re-

source costs required to produce products of the quality promised. The buyer's problem in this situation is to distinguish sellers who promise and intend to deliver high-quality goods from those making the same promise with no intent to incur the costs required to do so. The newer work shows that for Nelson-type experience goods, sellers can employ advertising as a credible signal of intent to deliver products with the qualities promised for the same reason that advertising signaled high quality in Nelson's analysis. A seller can recoup the costs of an expensive ad campaign only if consumers continue to buy its products in the future. A critical implication of this elaboration of Nelson's analysis to cover situations in which product quality is chosen by the seller is that costly signals can be employed to credibly communicate the signaler's intent, in addition to the status of innate traits over which the signaler has no control.

Other Costly Mechanisms for Demonstrating Intent

Advertising can function as a signal of intent to deliver a high-quality product because sellers who advertise have something to lose (profits on repeat sales) if they fail to deliver the quality levels expected by buyers. It is the fact that they have deliberately made themselves vulnerable to their customers' option to not buy their products again that makes the commitments credible. The economics literature identifies a number of mechanisms by which one party to an economic relationship can make credible a promise to provide goods or services to other parties in the future by voluntarily putting itself in a situation in which it has something to lose should it not follow through on its commitments.[20] For example, Williamson (1983) explains how mutual investments in the assets of a joint venture may make credible the venture partners' claimed commitments to the venture if the jointly held assets lose value if either partner fails to perform as promised. Camerer (1988) shows that there are situations in which a gift costly to the giver may credibly signal that the giver's intentions are honorable. An example is an engagement ring, legal ownership of which is transferred exclusively to the injured party should one member of an engaged couple unilaterally break off the engagement. In an influential article, Kreps (1990) shows how a number of prominent features of corporate culture, both as it relates to firm–employee and firm–customer relationships, may be explained by a signaling-type logic, whereby promises and implied commitments are made credible by the fact that there is a cost to be paid for breach of a commitment. Although the Camerer, Kreps, and Williamson articles are typically not included in reviews of the economics literature on signaling, the logic linking cost to credibility is the same.

Observance of Norms to Signal Commitment

Camerer's and Kreps' applications of the basic logic of signaling models to gifts and corporate culture suggest that this analytical perspective may be usefully applied to a variety of noneconomic behaviors and institutions. Posner (2000) developed this thesis in a recent book on law and social norms that has become a subject of debate among legal scholars.[21] Posner argues that a wide variety of behavioral norms can be interpreted as signals because their arbitrariness frequently makes it costly to observe them.[22] For example, a suit and tie are viewed as appropriate attire for men in most business meetings, although neither makes a tangible contribution to productivity, and meeting participants would be more comfortable in sweat shirts and jeans. Posner argues that the short-term personal costs of adhering to norms make their observance a reliable signal of willingness and intent to participate in a cooperative venture when the benefits of doing so can only be realized in the long term. Norms arise and are perpetuated because they are vehicles by which participants in social transactions can credibly demonstrate the depth of their commitments to each other or to a common enterprise.

COSTLY SIGNALS IN BIOLOGICAL CONTEXTS

In 1975, when economists were just beginning to explore the applications of signaling models, Zahavi (1975) offers a costly signaling explanation for a variety of previously puzzling features of animal mate-selection systems. Now commonly referred to as the *handicap principle*, Zahavi and others have since employed this basic signaling framework to explain prominent features of other animal communication systems, including those observed in predator–prey relationships. As in economics, once the intuitive logic of the signaling model was generally appreciated, an extensive literature developed that elaborated on the basic framework, tested its applicability to a variety of settings, and debated its meaning and inherent usefulness. Remaining doubts over the logical consistency and general applicability of the basic costly signaling argument were largely dispelled by two sophisticated mathematical analyses of evolutionary signaling games published by Graffen (1990a, 1990b). Maynard Smith's (1991) considerably simpler Phillip Sydney Game model was also important in demonstrating the formal support for the costly signaling hypothesis.

The economic signals discussed earlier allow signal recipients to distinguish between prospective transactional partners in two fundamentally distinct situations: situations in which transactional partners differ according to innate traits and capabilities they are powerless to change, as is the case with Spence's workers; and situations where transactional partners differ in their intent to follow through on a commitment (whether explicit or implied) to provide a product or service on terms acceptable to the signal recipient at some time in the future. Camerer's gift givers and advertisers who might promise products of one quality and deliver another provide examples of signals that operate in the second type of situation. In what follows, I refer to signals relating to innate traits and abilities as status signals and signals reflecting intent as intent signals. Although goals as diverse as corporate profits and marital bliss may be pursued by signalers and signal recipients in the interactions studied with the economic models, ensuring that one's genes are preserved in future generations is the overriding goal motivating the agents examined in the evolutionary analyses. Because the economic signals (or their levels) are in some sense chosen by the signalers, I refer to them as *volitional signals*. Volitional signals of status and intent are also examined in the animal signaling literature. In addition, the literature on animal signaling examines status signals with values that are not chosen by their bearers. I call these signals *nonvolitional signals*. I am aware of no obvious counterparts to nonvolitional signals in the economics literature on signaling.

Costly Signals in Mate-Selection Systems

Nonvolitional and both types of volitional signals have been described for the animal mate-selection systems that have been the subject of many, and perhaps most, of the signaling studies reported in the evolutionary communication literature. For the majority of species studied, females select among males who compete for their favors. Males may contribute to the viability of a female's offspring in two ways: through the quality of the inheritable traits they can pass on through their sperm and by sharing in the risks and burdens of parenting. For many species, males and females part ways almost immediately after mating, leaving the responsibility for caring for offspring entirely with the female. For these species, the hypothesis of fitness-based selection predicts that females will choose among males solely on the basis of criteria correlated with genetic superiority. Signals of male fitness that develop for these species should be status revelation signals.[23] For those species for which males may participate in parenting, signals indicating an intent to do so should also develop.

Lekking species, all of whose males do not contribute to parenting (Zahavi & Zahavi, 1997), provide numerous examples of volitional signals of status. Leks are common areas in which males of a species congregate to engage in competitive displays to attract females.[24] Insects, amphibians, fish, birds, and mammals are represented among the species with lek mating systems (Hauser, 1996). For many species of lekking birds, females choose males on the basis of their performance in elaborate and strenuous courtship dances, which may be performed daily for hours on end over a period of many days (Zahavi & Zahavi, 1997). Because males must decide how much time and energy to invest in competitive display, lekking displays are volitional signals.

Volitional Signals of Status. Lekking displays are obviously costly in terms of caloric expenditure and the loss of time that might otherwise be used to find and consume food. For prey species, visibility and vulnerability to predators are also increased. Physically more robust males should be better able to bear these display costs. If robustness is correlated with the genetic qualities females seek in a mate (this should be the case if there is a genetic component to fitness and fitness contributes to offspring viability), then from the perspective of females, lekking displays should be able to function as costly signals of the type described by Zahavi (1975). From the perspective of increasing the expected representation of their genes in future generations, males must consider the tradeoff between a higher likelihood of copulation if they invest more effort in display activities during any given display period and the higher likelihood of surviving to compete for females in the future if they display less vigorously in the current period. Grafen's (1990a) signaling model, which is a model of volitional signals, shows that fitter (higher quality) males should select higher levels for their signals. For lekking species, this means that fitter males should display more vigorously.

Field studies have provided considerable evidence consistent with the costly signaling hypothesis, but this is not universally the case. Unfortunately, difficulty controlling for the many factors that may vary in a natural setting makes it difficult to produce definitive evidence for the costly signaling model in this way. Ridley (2001) describes a study by Welch, Semlitsch, and Gerhardt (1998) that experimentally controlled for environmental factors that might vary in a natural setting and found that, for a particular species of tree frog, sperm from males with the longest mating calls produced genetically superior offspring. Long calls are more costly than short calls because they require a greater caloric expenditure and because frog predators use the calls to locate the callers. Females of this species show a distinct preference for males with longer calls.

Volitional Signals of Intent. For those species, including *Homo sapiens*, for which males may share the responsibilities of parenting, a female has a strong incentive to find out before mating whether a candidate male intends to stick around afterward. A male's intentions in this regard cannot be taken for granted because his interest in maximizing the representation of his genes in future generations might also be served by mating with as many females as possible—a strategy that is incompatible with staying home and taking care of the kids.[25] For these species, observable traits and behaviors predictive of the quality of care a prospective mate will provide future offspring should also be considered in females' deliberations over different males. Because the contribution a prospective mate will make to parenting can only be revealed after mating has occurred, we should expect female choice to give rise to intent revelation signals (in addition to status revelation signals) for males in these species. Dawkins (1976) lists nest building, feeding the female, and "a long engagement period" as demands females might make of males before consenting to mate with them. Camerer (1988) points out the obvious parallels between the nature of these premating demands described by Dawkins and the gifts given by humans that he describes as signals of intent. In each case, the resource commitment required is too great to have a positive payoff for a prospective partner that does not intend to be in the relationship for the long term.

Nonvolitional Status Signals. A wide variety of nonvolitional signals have also been identified as features of mate-selection systems. Common to these signals is a physical feature configured in a way that makes individual survival a more tenuous prospect than it would be if the feature were designed to maximize its bearer's ability to survive the physical challenges posed by the environment. The peacock's tail is probably the best known of these nonvolitional signals. It has even been used to illustrate communication tactics in a popular book on business strategy (Brandenberger & Nalebuff, 1996).

The size of the peacock's tail has been seen as a puzzle begging for explanation at least as far back as Darwin. A peacock's tail is so large that it makes flying more difficult, which makes peacocks more vulnerable to predation. In addition, the caloric expenditure associated with lugging their inordinately large tails around makes the task of finding the food required to keep avian body and soul together a more arduous and less certain enterprise than it would otherwise have to be. Because large tails reduce the odds of physical survival for their bearers, the peacock's tail (and analogous handicaps in other species) might be seen as a challenge to the Darwinian principle of survival of the fittest.

Darwin's pre-Mendelian explanation for the apparent anomaly of the peacock's tail was to point out that peacocks compete for mates as well as resources and to suggest that large tails were the evolutionary response to the clear preference of females for males with larger tails. Tail size thus contributes to male fitness by improving the odds of success in the competition for mates. Restated in terms of the modern debate over evolutionary processes in which an individual's success in the evolutionary game is measured by the prevalence of its genes in succeeding generations, selection should favor individuals with characteristics that make them more attractive to the opposite sex even if the consequence is a lower individual probability of survival as long as the tradeoff is not too steep.[26] Peahens can expect to have both male and female offspring (in most species, the ratio is 1:1), so once such a preference and its consequences for male mating success are established, selection will also favor peahens with a preference for males with longer tails. Because it is relative, not absolute, size that matters in the ranking of prospective mates, the constant culling of males with shorter tails will result in continual growth in tail size over time. Mathematical models have been developed that demonstrate that a female preference-induced arms race in male sexual adornments that reduce the odds of physical survival does not involve a logical inconsistency (Zahavi & Zahavi, 1997).

Zahavi and Zahavi criticize this explanation of handicaps for taking female preferences for male attributes as givens. Such preferences are presumably genetically determined, so the existence of a female preference for males with larger tails must also be explained in terms of selectionist pressures—that is, a preference for males with larger tails must confer an advantage on females possessing this preference over those for whom tail size is a matter of indifference at the time the preference is a novelty in the population. Zahavi and Zahavi propose that female preferences for prospective mates with the largest handicaps arise because the handicaps serve as reliable indicators of intrinsically desirable fitness characteristics a male might possess and pass on to a female's offspring. Physical strength and endurance, resistance to parasites and disease, tolerance for extremes in temperature, and an aptitude for finding the most nutritious types of food are examples of fitness characteristics females might find particularly important in a male. Males well endowed with these traits should also be able to more easily handle the physical demands of the types of handicapping features that are attractive to females.

As long as the handicap contributes less to the mortality rate of genetically fitter males than of less fit males, females can improve their odds of finding a high-quality mate by selecting the male with the largest handicap whenever faced with a choice. (They may also wait for a male with a larger handicap to turn up if only small handicap males are available initially.) Note that this does not mean that there are no high-fitness males with small handicaps, only that the odds of finding such a male are higher if mates are chosen from among the males with the largest handicaps. Thus, as with volitional signals, nonvolitional signals that persist are those for which the cost of the signal to the signaler is inversely correlated with the level of some other characteristic of the signaler that is particularly important to the signal recipient.

Costly Signals in Predator–Prey Relationships

The difference between volitional and nonvolitional signals also can be seen in relationships between predators and prey.

Volitional Signals. This discussion has focused on mate-selection systems to illustrate certain fundamental properties of the various signals discussed in the animal signaling literature. However, signaling explanations have been offered for a much broader range of behaviors and characteristics, including interesting features of predator–prey relationships. For example, prey birds have been observed to sing or whistle while fleeing a predator bird, such as a falcon, even though the energy required to do so could be employed to fly faster. The signaling explanation is that the quarry is signaling to the predator that is so confident of its ability to escape that it can sing while doing so. If this message is understood and believed, the pursuer will break off the chase, and both parties will save time and effort.

A similar explanation has been offered for the observation that some prey animals, such as deer, antelope, and gazelles, engage in stotting—making high and very conspicuous jumps—when fleeing predators, although doing so slows them down. Zahavi and Zahavi (1997) cite studies of gazelles pursued by hyenas and African wild dogs that found that not all gazelles stott, and that the dogs and hyenas concentrated their pursuits on individuals who either did not stott or stotted very little. The signaling explanation is that predators "understand" that the cost of stotting is too high in speed and endurance sacrificed for the weaker and slower gazelles.

Nonvolitional Signals. The distinctive markings and coloration of butterflies such as monarchs, which taste nasty to birds, have also been explained as a costly signal.[27] Distinctive coloration and markings are an advantage for monarchs and other "nasties" because they make it easier for birds to remember what type of butterfly not to attack after having an unpleasant experience with one of them. (Animals that taste terrible to their predators are commonly referred to as nasties.) Easily recognized and remembered features would be a disadvantage for butterflies that taste good to birds because they would reinforce the memory of a positive experience. The same explanation has been offered for the fact that poisonous frog species are often brightly colored and highly visible to animals that prey on frogs. The observation that nasty species frequently look like other nasties is explained by the informational advantage of sharing a common signal.[28]

VOLITION AND COMMUNICATION

Volitional and nonvolitional signals are both informative, but only volitional signals are communicative. Volitional signals reflect choices made by one behavioral agent trying to influence another. Yet it is nature, not a behavioral agent, that selects nonvolitional signals. Observing the size of a peacock's tail, a peahen learns something about the severity of the survival test he had to pass just to have the opportunity to spread his tail before her. With this information she can make a more informed choice among prospective mates, employing a decision rule that embeds an understanding of the effect of signal cost on the survival rates of males of varying quality. Yet no such understanding is required of the peacocks vying for her favors, who have no choice but to compete with the tails nature has given them.

Because the senders of volitional signals choose the levels of their signals (how much time to spend courting a female, how much to spend advertising a product), there must be some sense in which they also understand the logic of costly signals if these choices are to serve their interests—which must be the case if selection preserves fitter decision rules over the long run. Consider, for example, a male bird engaged in a courtship dance. The decision of how long and hard to dance should reflect the dual effects of dancing on his survival odds (increased vulnerability to predators and higher caloric needs) and his chances of attracting a mate. This second consideration means that a male's decision rule must incorporate an understanding of the decision rule employed by females in selecting mates, which in turn, if it is to serve their interests in propagation, must reflect an understanding of the decision rules employed by males. This is a concrete example of the general model of conditional expectations communication depicted in Fig. 3.3.

Although the same inferential logic of costly signals underlies the responses to both volitional and nonvolitional signals, the shift from nonvolitional to volitional signals is a shift to a much more complex set of interactions with coordinated decision rules. This does not mean, however, that volitional signaling is in any sense less deterministic than is nonvolitional signaling. It would be hard to argue, for example, that insect and amphibian lekking displays are the products of conscious deliberation. Rather, lekking behaviors reflect instinctual decision rules that are just as much the products of natural selection as are the nonvolitional signals of other species. As Spence (1973) points out, there is no reason to assume that the decision rules underlying economic signals reflect a conscious understanding of how these signals actually work. Signal senders and receivers simply have to recognize their effects. Volition cannot be treated as synonymous with free will in this analysis of signaling or for conditional expectations communication more generally.

BROADER APPLICATIONS OF CEC

In the section introducing the basic CEC model, I suggested that this framework might be used to model a variety of communication practices, including conversations that progress through multiple stages. The discussion since has focused largely on various types of signaling to take advantage of the findings of two rich literatures on the topic developed by economists and evolutionary biologists. This section returns to the broader theme to (a) point to other work that incorporates conditional expectations, (b) provide examples of activities that normally might not be viewed as communicative but can be interpreted as such from a CEC perspective, and (c) offer a few illustrative examples of practices commonly recognized as communication that might fruitfully be analyzed in terms of the CEC model.

Credible Communication Without Costly Signals

Costly signals address the problem of assuring message credibility when the possibility that one communicating agent may benefit from misleading another cannot be dismissed out of hand. This suggests that there may be situations in which a mutuality of interest is so obvious that a message receiver can unerringly infer that it is in the best interest of the sender to represent itself honestly. In such situations, the message or claim would be the only element of the environment modified by the sender. For example, a mother grouse's cry of warning may be accepted at face value by her chicks foraging nearby because her genetic interest in their safety aligns her self-interest with theirs. The conditions under which costless messages can be accepted as credible has been examined in the economics literature on "cheap talk."[29] Similar questions have also been addressed in the literature on the evolution of biological signals (Maynard Smith, 1994).

Even when mutuality of interest is not obvious, a costless signal may be interpreted as reliable if it is associated with other costs in a relationship. For example, Johnstone and Norris (1993) describe an equilibrium in which the size of a bird's badge of status (a patch of brightly colored plumage) may be interpreted by other birds of the same species as an index of aggressiveness if aggression is costly, even though the badge is not. The Moore et al. (1997) study of dominance and mate selection in cockroaches suggests that chemical scents may play the same role.

Braces, Extracurricular Activities,
and Competition for Resources

Most behaviors can be explained by both the immediate goals they promote and in terms of the workings of a more complex process that makes those goals matters of importance to the agents pursuing them. This subsection provides three examples of activities not normally thought of as communicative for which CEC appears to offer insightful process-level interpretations to illustrate the wide range of activities that might be analyzed from a communication perspective.

Braces. People exhibiting strong left–right body symmetry are generally perceived as more attractive than less symmetric individuals. Ridley (2001), among others, suggests that this preference has evolved because symmetry and regularity of features are reliable indicators of genetic fitness, by which he means a low incidence of harmful genetic mutations in an organism.[30] Although various compensation mechanisms allow organisms to survive with a variety of genetic defects, symmetry is still correlated with good health. In other words, symmetry is a nonvolitional signal of genetic fitness. Being nonvolitional, symmetry is an informative, but not a communicative, signal in a state of nature. Zahavi (1993) argues that body symmetry is a costly signal of fitness because in responding to stress the body draws on resources that might otherwise be used to coordinate symmetrical development. If we accept this signaling explanation for a preference for body symmetry, this and other preferences underlying generally accepted standards of beauty and physical attractiveness are deeply rooted in the genome and can be expected to persist for many generations even if the environmental exigencies that produced them are eliminated by advancing technology.

Dental braces are employed both to improve a person's bite and make the distribution of teeth in the mouth appear more regular and symmetric. In an environment in which life is a more tenuous proposition than it is in modern industrial societies, it seems reasonable that the appearance of teeth might serve as one among a number of indicators of fitness. When used for their long-term cosmetic effects, braces (and other cosmetic medical procedures) are attempts to acquire through financial means the advantages of a signal that nature did not supply. Parents may explain their often substantial expenditures on braces in terms of their long-term contributions to the happiness of their children, but at the same time they are increasing the odds that their children will find mates and keep their parents' genes in the collective gene pool for at least one more generation. From this perspective, braces and cosmetic medical procedures in general are fundamentally communicative acts. In the case of parents paying for their children's braces, their children's teeth are the modified elements of the environment, and their children's prospective mates are the agents whose behaviors they seek to influence.

It is worth noting that as an attempt to procure the benefits of a signal whose natural interpretation does not apply to their wearers, braces are also a deceptive communicative act, although the intent to deceive may not be conscious. The issue of deceptive communication is addressed in the theoretical literature on signaling in biological systems (see e.g., Johnstone & Grafen, 1993) and in a number of studies of deceptive communication practices employed by a variety of animal species (these studies are reviewed in Hauser, 1996, chap. 7). Key questions in the analysis of deceptive communication practices are how large a population of deceptive signals can be tolerated before the honest signaling system it exploits breaks down (revised conditional expectations no longer justify the desired response) and whether a stable equilibrium is reached prior to this point.

My own casual observation of changing social attitudes regarding braces suggests that braces have acquired a social signaling value independent of their effects on the long-term beauty of their wearers' smiles. At one time braces were sources of embarrassment hidden behind closed lips—perhaps a tacit acknowledgment that their wearer was attempting to procure a signal to which she was not rightfully entitled or perhaps because braces call attention to the state of the wearer's natural teeth. Yet braces also demonstrate that their wearer (or her parents) has the financial wherewithal to pay for them. Today most children do not seem to be self-conscious about their braces, and in fact often draw attention to them with multicolored wires and bright paste gems. The fact that most children's braces are off long before the mate-selection process begins in earnest may account for what seems to be the dual use of braces to acquire a signal of genetic fitness and as a signal of economic status. Although the empirical foundation for this interpretation of expenditures on cosmetic denture work is admittedly casual, I believe it faithfully illustrates a way in which genetic and social processes generating conditional expectations may interweave in the development of signaling systems and communication practices more generally.

Extracurricular Activities. Whereas grades and test scores were once largely determinative of college admissions decisions, today most high school students believe that a strong record of participation in extracurricular activities substantially improves their chances for admission to a top college or university. The growing importance of extracurricular activities in college admissions has coincided with a substantial growth in the number of students seeking admission to the top schools and grade inflation throughout the educational system. Grade inflation narrows the grade point differences among top students. Combined with growing numbers of applications, it has placed admissions officials at the more selective schools in the unenviable position of having to choose among large numbers of strong applicants with virtually identical scholastic records. Grades, by themselves, should therefore be less predictive of ability differences among top applicants than they once were.

High grades combined with an impediment to achieving them is a different matter, however. Admissions officials might reasonably infer that students who make high grades while participating in enough extracurricular activities to severely restrict the time available for study are better equipped (on average) to meet the challenges of a college curriculum. (Note that this *understanding* could just as easily be expressed in terms of the predictive value of evidence of broad or well-rounded interests. Recognition of the correlation, not an understanding of the causal relationship, is all that is required for the signal to work.) This is not to say that students with high grades and low levels of participation in extracurricular activities will always have lower abilities than students with comparable grades and high participation levels, only that the odds of admitting high-ability students are greater for the pool of students with strong records of extracurricular interests.

The admissions officer's situation is similar to that of the peahen selecting among peacocks with different sized tails. Although any individual peacock with a short tail may be just as robust as any individual peacock with a large tail, the lower survival rate of less robust males with large tails means that the odds of getting a robust male are higher if the female selects among those with large tails. Similarly, the negative impact of a heavy load of extracurricular activities on grades should be greater for those students with less innate academic ability. Recognizing the signaling value of extracurricular activities, students (and their parents) take pains to build them into high school schedules even when the students' innate interests alone might not lead them to do so. Thus, what was probably once a purely informative signal is now communicative as well.[31]

Competition for Resources. It was argued previously that although the length of an individual peacock's tail is informative to peahens, it is not communicative because tail length is determined by a peacock's genes and the nutritional resources accessible to it in its environment—neither which are under its control. On the other hand, if one peacock tried to deny other peacocks access to food sources even though there was no personal nutritional gain to doing so, that would be communication if the end served was the mate-selection advantage realized from having competitors with shorter tails. I raise this possibility not from personal knowledge of the manner in which peacocks compete for food, but to illustrate the range of activities that might be analyzed in terms of CEC. The same perspective applied to the antics of humans suggests that if the trappings of wealth are a source of advantage in competing for desirable partners (whether sexual, social, or economic), then the signaling consequences alone should lead the rich to oppose and the poor to favor government policies that redistribute wealth—independent of the effects of redistribution on consumption opportunities.

CEC Applications to Familiar Communication Activities

Many other applications of the CEC framework are possible, including analysis of familiar communication practices that are considerably more complicated than those investigated in the signaling literature. I close this section with a few suggestive examples.

Somewhat more complicated than the signaling relationships examined in this chapter would be a situation in which one agent's provision of information to another achieves its objectives by modifying the second agent's response to information from yet another source (which may or may not be another agent) by causing the second agent to revise the probabilities assigned to various behavior–outcome combinations and/or the importance weights assigned to outcomes. This would appear to be the logic underlying the use of negative advertisements on behalf of products and politicians. Ads that dispute claims made for a competing product (or politician) are broadcast to increase the probability that buyers (or voters) will support the product (or politician) supplying the ad. The negative ads are effective only to the extent that they cause listeners to revise the probabilities and/or importance weights assigned to an opponent's claims.[32]

Arguments based on deductive logic would appear to satisfy the CEC definition of a communicative act as long as it is easier to construct convincing logical arguments for true claims than for false claims. Logical arguments also expose their purveyor to the risk of being exposed as fraudulent on purely logical grounds. Finally, if we allow outcomes to include decisions of whether to continue communicative interactions, it should be possible to model and analyze multistage conversations from a CEC perspective.

BIOTECHNOLOGY AND HUMAN COMMUNICATION

I now return to the question raised at the beginning of this chapter: Can biotechnology change the nature of human communication? If by the nature of communication, we mean the fundamental logic governing the development of communication processes in selectionist systems, I believe the answer has to be no. I argued before that communication practices based on conditional expectations are products of selectionist pressures. Unless biotechnology eliminates selectionist pressures entirely (and short of eliminating the biological and social bases for mutation and change at the level of the individual, I do not see how it can), communication practices based on conditional expectations are inevitable. There are, however, several interesting ways in which biotechnology might influence the development of CEC practices.

Biotechnology can be employed for two general types of genetic interventions: prefertilization alterations of the DNA in gametes before they fuse with their sexual complements to create new organisms, and postfertilization modification of the genetic material of somatic cells in a developing or already developed organism. Interventions of the second type might be viewed as biotechnological analogues of dental braces. Two types of gene therapy for Type I diabetes that are currently being researched are examples of the second type of genetic intervention. Type I diabetes is an autoimmune disease with genetic origins that destroys the insulin-producing beta cells of the pancreas. One type of gene therapy being investigated for Type I diabetes would implant in a patient's body insulin-producing cells that have been genetically engineered to prevent either rejection or autoimmune destruction by their new host. The second type of gene therapy under investigation would deliver new genetic material to non-beta cells within the patient to engineer them to produce insulin as a substitute for the insulin that would have been produced by the beta cells destroyed by the patient's immune system. Either approach, if successful, would eliminate the symptoms of a genetic defect. In theory, there is no reason that similar gene therapies could not be developed to accomplish the work of dental braces or other types of cosmetic medical procedures. Yet like braces, these types of genetic interventions would contribute nothing to the genetic fitness of their beneficiaries, whose likelihood of passing their genetic defects on to their prospective offspring would be unaltered. If good health and physical attractiveness are interpreted as signals of genetic fitness, they are misleading signals if acquired as consequences of postfertilization gene therapies.

Prefertilization gene therapies would be another matter. Genetic modifications to gametes are inheritable. Thus, the cosmetic consequences of therapeutic interventions to improve the DNA of individuals yet to be conceived would be reliable signals of fitness to potential future mates.

This discussion of the extent to which the cosmetic consequences of gene therapy might or might not be interpreted as misleading signals of genetic fitness has taken preferences for symmetry and other physical manifestations of health and genetic fitness as givens. Yet the genetic bases for these preferences are at least in principle modifiable by the same biotechnologies that might be employed to cure medical problems with genetic origins. In a world where cheap and generally available gene therapies cured all genetic defects, it is not clear that such preferences would serve any adaptive purpose. Nevertheless, as long as they persist, the pressures to employ biotechnologies for cosmetic purposes would be difficult to resist. The saving in resources and social conflict that could be avoided by eliminating preferences for physical traits in other individuals that function as signals of genetic fitness would probably be enormous, although the genetic bases for other behavioral tendencies such as aggression might be more obvious candidates for modification. But how much individuality would be sacrificed in the bargain? And what philosophical principles might inform a discussion of the ethics of modifying the genetic bases for preferences and other behavioral tendencies that are fundamental elements of personality?

ENDNOTES

1. I want to thank editor Sandra Braman for encouraging me to write this chapter and for suggestions provided in response to earlier drafts of the work. This chapter builds on research conducted while I was the 1996–1997 Van Zelst Research Professor of Communication at Northwestern University.

2. If, as many now believe, preferences evolve to promote fitness, preferences for certain attributes in prospective mates are driven by the more primary (although perhaps unconscious) motive of positively influencing the traits, and thereby the fitness, of future offspring. Biotechnology offers a more direct means to the same end. If the underlying motives are the same, it is not altogether obvious why the biotech approach is in any way less ethical or noble than the old-fashioned method of influencing the fitness of one's offspring.

3. Russell (2002) offers a review of recent research on this subject for the general reader.

4. I call the provision of such information, whether perceived by its intended recipient or not, a *communicative act*. There is communication when the agent to whom the information is directed perceives it.

5. Success in a biological system is typically defined in terms of perpetuation of a gene or collection of genes. Thus, responses to environmental variation that promote the biological success of an individual may allow for altruistic self-sacrifice when doing so increases the odds of survival of other individuals sharing the same gene or genes.

6. Assigned probabilities are subjective in the sense that they are an individual organism's estimate of the probabilities that specific outcomes will be associated with specific behaviors. Subjective probabilities could be consciously calculated and explicit in the case or humans, or they could be implicit in the decision algorithms employed in selecting among behavioral options for both human and nonhuman actors.

7. I take some literary liberty here in ascribing goals (or purpose) to individual organisms. *Survival of the fittest* means nothing more than, on average, over time, genes that produce individuals with inherent advantages in reproducing themselves will displace genes that produce individuals that are less advantaged (less fit) in this way. Genetically influenced behaviors that promote fitness in this sense will make individuals employing them appear to be motivated by a desire to procreate. Selection for fitness in a biological setting will produce individual organisms that behave as if they are trying to perpetuate their genes or trying to accomplish intermediate goals that promote the same end.

8. See Hauser (1996) for a list of other definitions of communication that, like this one, emphasize an intent to influence the behavior of another organism.

9. The first agent may be pursuing objectives for which further interaction with the second agent is not required. For example, the first agent's ultimate interest may be in the consequences of the second agent's response to the communicative act for a third agent, as would be the case with a government campaign urging parents to tell their children not to use drugs.

10. The possibility that communicating agents might be plants was mentioned in the introduction. Plants may also communicate with insects in a manner consistent with CEC communication. Milet (1997) reviews research showing that tobacco plants attacked by insects release chemicals that attract other insects that prey on the bugs chewing on them. The chemical signals appear to be specific to the type of insect attacking the tobacco plants. To be reliable, such signals should be costly to plants, and thus restricted to times of distress. If not, the signaling system would be susceptible to invasion by mutant plants producing the chemical attractants all the time as a preemptive defense. Yet if this were to happen, the chemicals would no longer be informative to the predators they were meant to attract. Predator insects that continued to respond to the chemical attractant would then be displaced by genetic variants of their species who did not.

11. Although I have not seen this argument made elsewhere, it seems likely that genetic selection and social learning interact in the formation of certain, and perhaps most, types of expectations. Genetic selection could determine expectational priors that are then modified in response to experience in environments that may fluctuate over time.

12. Female choice is common in the creation of mating pairs by nonhuman animals.

13. In the economics literature, this would be referred to as a fair bet because the sum of the probability weighted outcomes equals the amount paid to take the gamble.

14. Recognizing that the prize goes to the one who runs the longest, the teaching pros might be expected to observe each other during the course of the run to gauge the energy reserves of their competitors because dropping out early would be the optimal strategy in a competition with a determined and clearly superior opponent. The fact that the outcome now reflects the participants' assessments of each other opens up possibilities for strategic bluffing—a matter that has been extensively discussed in the animal communication literature, but has received little attention in the economics literature. Zahavi and Zahavi (1997) argue that strategic bluffing in competitive situations is also a competition in the provision of costly signals because the contestants look for indicators of relative capability that are difficult (or costly) to fake.

15. See also his book, *Market Signaling* (1974).

16. Spence shared the prize with George Akerlof and Joseph Stiglitz.

17. Spence pointed out that within the feasible range there may be an infinite number of equilibrium values for the signal.

18. In the extreme, low-quality sellers might drive high-quality sellers from the market if consumers cannot reliably distinguish between them.

19. The ability of sellers of high-quality products to cover the costs of an ad campaign increases relative to the ability of sellers of low-quality products to do so the larger is the ratio of repeat to initial sales and decreases the larger are any production cost advantages realized by sellers offering a lower quality product.

20. Goodwin (2001) makes a similar point in arguing that the source of Cicero's authority in his defense of Sulla was his audience's awareness of the damage his valuable reputation would suffer if it were discovered later that he had misled them.

21. See for example, Mahoney's (2001) review and critique of Posner's book.

22. Kreps makes a similar point with respect to corporate culture.

23. With female choice and honest signals, males are fit to the extent that they contribute to female fitness.

24. The similarity to high school football fields and basketball arenas is hard to ignore.

25. Males have a parallel concern with the faithfulness of their mates because females may pursue a "best-of-both-worlds" strategy of nesting with one male and copulating with higher status males on the sly. Genetic tests of nestmates have shown that it is not uncommon for multiple fathers to be represented in a nest.

26. If p is the probability that an individual male lives long enough to reach sexual maturity and q is the probability of successfully mating having reached that age, over time selection will favor males for which pq is highest. If larger tails produce a percentage increase in q that is larger than the percentage by which they reduce p, selection will favor larger tails.

27. Monarchs acquire their nasty flavor from chemicals in the milkweed plants on which they feed.

28. Although what might be called *signal parasitism* is also possible, in which non-nasty species take advantage of what predators know about nasty species by evolving similar markings, as appears to be the case with the monarch mimic, the viceroy.

29. See for example, Farrel (1987) and Farrel and Rabin (1993, 1996). Bagwell and Ramey (1993) show that there are conditions that support cheap talk equilibria for advertising claims.

30. According to Ridley, an unavoidable consequence of the number of genes required to make complex organisms is that most individuals of a complex species will carry a substantial number of harmful, although nonfatal, genetic mutations.

31. Participation in extracurricular activities is also a volitional signal because its level is selected by the signaling agent. It is probably natural for volitional signals that start out informative to evolve into communicative signals unless nonsignaling considerations are largely determinative of their levels.

32. Competitive strategies targeting claims made by rivals apparently are not unique to humans. Male bower birds build elaborate display courts (bowers) that are examined by females as part of the mate-selection process. Apparently recognizing that it is the relative, not absolute, size of the bower that counts, male bower birds also attempt to damage each other's bowers (Andersson, 1991).

4

"Are Facts Not Flowers?":
Facticity and Genetic Information

Sandra Braman
University of Wisconsin–Milwaukee

Are facts not flowers?
and flowers facts?

—William Carlos Williams

Ever since Locke (1690/1998) introduced the notion of the *fact* in 1690, mediated human communication has revolved around facticity—the claim that a given datum or narrative does or does not refer to empirical reality. Distinctions between fact and fiction, fact and opinion, fact and falsehood, and fact and error are important for legal, economic, and often social treatment of information and those who produce it. Genres became distinguished along these lines so that journalism and history, for example, are presented as factual, whereas fiction is not. Adherence to specific practices and procedures for the production of fact are the core of professionalism in law, journalism, and science. Institutions with the capacity to "certify" data and narratives as factual have developed, always associated with governance and power.

Facticity has long been a subject of communication research in areas as diverse as studies of news production, sense making, the social construction of reality, and persuasion. Because technological change alters the practices by which facts are produced and the institutions by which they are certified, with each innovation in information and communication technologies new questions about the nature of facticity and its production arise and old ones must be revisited. The printing technologies of Locke's time presented a challenge to facticity because print publications offered the possibility of immediacy but did not guarantee it. As a result, "with the beginnings of the report of recent events came the problem of proving the truth of that report" (Davis, 1983, p. 70), a difficulty repeated on a different time scale with the telegraph (Headrick, 1990) and, today, the internet. In the same vein, the electrification of communication technologies in the late 19th and early 20th centuries made possible new forms of mass media and, consequently, genre shifts that led Robert Park to comment in 1940 on the rise of a "specious present."

Since that time, commodification of the fact (Lyotard, 1984), the increasing ease with which seeming facts can be manipulated (Enzensberger, 1974; McLuhan, 1968; Zavarzadeh, 1976), and growing sophistication regarding the interaction between facticity and power (Foucault, 1972) have made facticity a matter of focal interest. With meta-technologies, the expansion of the degrees of freedom with which information can be processed undermines facticity because the number of steps in the causal chain through which facts are produced has multiplied, interactions between facticity at different levels of granularity must be taken into account, and there is a proliferation of facts in a Lockean sense that must be taken into account in the formation of meaning. Just as the scale, scope, and ubiquity of human communication has been transformed in the digital environment, so the problems of producing, evaluating, and sustaining facticity have become more complex.

Genetic information of course differs from the facts of human communication because it is a set of instructions—like a computer program—rather than an assertion, but the effects of both types of facts we now know vary according to the receiver, context, and environment in which meaning, or form, is made. Genetic information raises issues of facticity at three levels: First order issues arise when determining the facts of genetic information itself, including questions of identification, quantity, and effect. The institutions that certify genetic information in various ways both demand facticity and constrain the ways in which representations can be undertaken, producing second order issues. Discourse about genetic information in society at large, often with persuasive intent, raises third order facticity issues in familiar genre form.

As with the facts of human communication, the facticity of genetic information matters because it has economic, social, and cultural implications. There are political implications as well because transfers of genetic information led to a radical restructuring of the global division of labor and control over it following the "Columbian Explosion" (Crosby, 1972, 1994), and practices based on misrepresentations of genetic information can result in environmental and/or health disasters. Today's ability to alter the genetic makeup of people—to change the facts, so to speak—contributes to the sense that we are entering the "posthuman" era so much discussed by analysts of the effects of digital information technologies (see e.g., Dewdney, 1998; Hayles, 1999). Both types of meta-technologies—biotechnology and digital information technologies—affect confidence in a fixed empirical reality and/or our ability to perceive it, habits and norms of fact production, the nature of the institutions that certify facticity, and the possibility of asserting property rights in facts.

The "semiotization" of nature (Hoffmeyer, 1997) began in the early 20th century when biologists began to explore the ways in which organisms perceive and act on their environments in ways describable as communicative, and it was extended when the discovery of DNA suggested it was possible to understand biochemical processes in the same way. There are of course limits to the extent to which metaphors can legitimately be transferred in either direction (see Ritchie's examination of this problem in chap. 2, this volume). Looking at the facts of genetic information and of human communication through a common lens, however, makes it possible to learn from the history and sociology of one type of meta-technology ideas that can usefully be applied to the other. Wildman (chap. 3, this volume) examines one such communicative feature—the role of conditional expectations. This chapter explores another, suggesting a typology of types of facticity evident in the analysis of genetic information that can be applied to the study of human communication.

The problem of facticity does not exhaust the ways in which biotechnology is intertwined with the sociology of knowledge. Development of the field has disturbed historical distinctions between types of scientific specialization; created new relationships among government, industry, and academia (Benson & Shaw, 1992; Hadwiger, 1982; Kloppenburg, 1988; Lazonick, 1991; Wiegele, 1991); and provoked new theories about the development of disciplines and their life cycles (Latour, 1988; Vernon, 1990). Indeed, the methodological revolution after the 1880s that led not only to contemporary biotechnology, but also to the use of research-directed experiments across the social and physical sciences, was driven at least in part by the very desire to raise respect for the field of biology (Harwood, 1993; Pearton, 1984). The sociology of knowledge has been useful in the diffusion of knowledge needed to use the products of biotechnology (McCorkle, 1989). The requirements—or alleged requirements—for knowledge building in biotechnology has led to changes in the law and government institutions (Kloppenburg & Kleinman, 1988; Pearton, 1984). Lievrouw (chap. 6, this volume), Murdock (chap. 9, this volume), and May (chap. 5, this volume) address some of these matters further.

THE GENE AS FICTION VERSUS GENETIC
INFORMATION AS FACT

Few concepts in the history of the physical sciences have been as problematic as that of the gene (Goonatilake, 1991; Nelkin & Lindee, 1995). It is paradoxical that while the gene began as a linguistic fiction and continues to be so, it functions in popular culture as a secular equivalent of religious truth in ways that have remained remarkably consistent since the early 20th century. Although genetic information is constantly mutable and mutating, in popular culture the gene is treated as incapable of deception and as the locus of authenticity.

The term *gene* was coined by Danish geneticist Wilhelm Johannsen in 1909 to describe a presumed cellular entity capable of producing a particular trait. The notion was inspired by German physiologist and geneticist Hugo DeVries, who used the word "pangenes"—itself derived from Darwin's concept of "pangenesis"—to refer to the origins of biological variation. For the first generation of experimental geneticists in the early 20th century a gene was, in practice, a physical trait such as wing shape or eye color which seemed to derive from a substrate of hereditary material, the actual constitution and functioning of which were at the time unknown. More recently, increasing knowledge about the gene as a molecular entity has clarified its physical form but complicated its biological meaning, not in small part because it can be described concurrently in morphological, physiological, genetical, and molecular terms (Bent et al., 1987).

The confusions are two, both deriving from one of the most ancient of philosophical problems—the difficulty of understanding the relationship between the intangible and tangible. First, in the popular imagination, the physical material in which genetic information is embedded—DNA—is often equated with the information embedded within the DNA itself. This is akin to mistaking the physical object of a book for the information contained within it. Second, the distinction between the genotype (the genetic information of which an organism is comprised) and phenotype (the manifestation of that genetic information in the material world) is often conflated. The phenotype results from interactions of the genotype with particular environments. Indeed, within any one species there may be thousands of the genetically variable populations derived from differential responses to environmental conditions known as landraces. Although it is tempting to think of the genotype as an ideal Platonic form, it is more accurately described as a potential for a myriad of possible forms out of which one will be selected by the environment and the history of an individual within it. There are important implications of this difference in the social world: It is one reason that, for example, genetic testing in the workplace should not be grounds for excluding certain workers from particular jobs because there is no way to know whether a genotype will be expressed phenotypically in any particular individual and/or context. There is in reality no single normal genotype, but rather an entire spectrum of genotypes should be considered normal (Suzuki & Knudtson, 1989). It is the genotype that is the genetic information discussed here.

Because of these two confusions, the word "gene" no longer has meaning for molecular biologists. "But," note Hubbard and Wald (1993), "since genes remain very much a part of the science of genetics, as well as of the culture at large, experiments with DNA get communicated in the language of genes" (p. 43). The result is ambiguity. Although DNA, the genotype, and the phenotype are so intertwined that it is not always clear which is being discussed, the actual complexity of genetic information often gets translated—both within the scientific community and without—into a machine-like metaphor that emphasizes the ability of humans to control it (Haraway, 1976; Nelkin & Lindee, 1995; Woodward, 1994). Even when there is no such confusion regarding genetic information, however, there are first order, second order, and third order facticity issues.

GENETIC INFORMATION AND FIRST ORDER FACTICITY

First order facticity issues involve the identification and description of information. They fall into two categories: "natural" falsification and human falsification. The themes are familiar because first order facticity issues of genetic information deal with two relationships: between sign and referent, and between stability and change.

Natural Influences on the Facticity of Genetic Information

The facticity of genetic information can be affected in natural ways through environmental influence, the presence of empty seed husks, or of nonsense, decay, genetic drift, and masking. This type of facticity issue has importance in the marketplace because buyers want to know what they are getting. In addition, in some cultures and/or periods of cultural history, the purity of species has been a moral issue (Harwood, 1993; Wiegele, 1991).

Environmental Influence. The genotype/phenotype relationship inevitably leads to a situation in which what one thought was a genetic fact may not be so. Efforts to prevent the phenotype from wandering too far from the genotype are essentially hopeless, although those responsible for maintaining genebanks put a great deal of effort into trying to maintain purity for purposes of quality control. However, manipulating the environmental factors that influence the phenotype can be a biotechnology technique in itself. Following a classic 1927 demonstration of artificial mutation by x-rays, Germans and Russians tried to target—or intentionally induce—particular types of mutation (Fleming, 1968) and following WWII, the Japanese successfully used radioactivity to create genetically mutated organisms for commercial uses such as industrial fermentation (Krimsky, 1991). The meaning of human communication of course also differs significantly from context to context. Thus the focus in recent research has been on reception practices rather than content alone (see e.g., Liebes & Katz, 1999).

Decay. Genetic information can be "archived" in genebanks, but without regeneration (planting and reharvesting seeds or inducing reproduction in animals), the information held decays over time. Each type of genetic information has its own lifespan—wheat must be reproduced every 3 years, alfalfa every 10, and potatoes yearly. Long-term storage holds seeds for 50 to 100 years, but this requires moisture-controlled conditions at temperatures below freezing (at the highly sophisticated genebank in Ft. Collins, Colorado, many seeds lie in liquid nitrogen at –273 degrees F to minimize decay). Only a small proportion of the world's genebanks have such facilities. Some of the world's most important archives of genetic information, such as the Vavilov collection in Russia, hold seeds at room temperature ("Needed," 1994; Strobel, 1993). Even regeneration does not prevent decay, however, for it never exactly replicates the genetic information involved because of genetic drift and alterations in the phenotype (Konopka & Hanson, 1985). Some "falsification" also results because people automatically choose the best plants in every regrowing cycle, gradually adapting the original genetic composition to the new environment and favoring certain characteristics over others even when not for survival value (Leeuwis, 1993).

Thus just as there is the problem of conservation for the materials of human communication, whether in paper or electronic form, so it is difficult to conserve genetic information. As with human communication, where errors are seemingly inevitable with most modes of copying, each time there is a reproduction of genetic information the process itself introduces error. In human communication, the problem of decay has gotten worse with each technological innovation—whereas symbols chiseled into stone thousands of years ago remain legible, electronic storage media must be refreshed every couple of years. Recently software that was developed to study the evolution of biological organisms has been applied to the analysis of changes in texts that result from error occurring during copying, leading to new interpretations of Chaucer (Brainerd, 1998).

Empty Seed Husks. The effort to quantify genetic information can founder when what is expected to be a physical embodiment of genetic information does not actually include that information, as when a seed husk is empty. In the case of seeds, the problem of distinguishing between seeds and seed-like structures is frequent enough that it is of keen interest to those in agriculture who are, for example, concerned about such matters as yield in seeds per acre. Estimating the number of seeds is also important to managers of genebanks because it determines whether material can go into long-term storage without first being reproduced, distribution policy, the number of seeds available for testing, how much space is required for storage, and the date the next generation must be produced. In some cases, a purity analysis is needed simply for counting purposes. One procedure developed to cope with this problem is to divide the total weight of seeds by the estimated mean seed weight as established by the International Seed Testing Association (ISTA) to get what is considered to be a reliable figure for the actual number of seeds (Konopka & Hanson, 1985).

Similar problems arise in the struggle to develop research methods that are both valid and reliable for the study of human communication. Formulaic techniques are also used here to address those problems.

Nonsense. In human cultures, nonsense fulfills the function of play. Some genetic information is described as nonsense—"junk DNA"—because it is not yet understood scientifically (Krimsky, 1991; Nelkin & Lindee, 1995). In the case of genetic information, such material actually marks the bounds of human knowledge and is not a matter of facticity at all. Indeed, many of the most important evolutionary features do not arise as adaptations, but as biological, social, or cultural cooption of structural byproducts—"spandrels"—thrown off from adaptive change for new purposes. Many even believe that reading and writing evolved in this way (Gould, 1997).

Change of Referent. In human communication, a vast amount of effort has been devoted to explicit discussion of the relations between signs and referents—definitions. In many cases there is more than one definition for a word or phrase, but when that happens each referent is separately identified in comprehensive dictionaries. Such an accomplishment is many years away with genetic information or may never be possible. Identical sequences of genes at different points on the genome can have different biological meanings, and the same genes can have different effects in different organisms from the same species because of the complexity of interactions among genes, and between genes and the environment.

Genetic Drift. Neither genotype nor phenotype remains stable over time as a result of genetic drift, the natural introduction of changes in genetic information in the form of mutation. This is a difficulty for intellectual property rights in the same way that asserting those rights in constantly changing digital texts is problematic. It also presents a problem for economists, who define goods as stable in form across time and space. It was for just this reason that the U.S. Department of Agriculture resisted the patenting of sexually reproducing plants for many years. Although the Plant Patent Act of 1930 did mark a significant change in the ability to patent asexually reproducing plants, it took several decades before those that sexually reproduce became patentable because the degree of genetic drift of such plants is higher. When they did, with the Plant Variety Protection Act of 1970, it was not because of a belief that sexually reproducing plants were any more genetically stable than they had been, but because the PTO stopped requiring precise descriptions of specific individual examples of the genes for which a patent was sought in favor of requiring only a description of population parameters (Bent et al., 1987; Kloppenburg, 1988; Krimsky, 1991).

The parallel in human communication would be natural changes in meaning that can, over time, lead readers or viewers to misinterpret a text created in an earlier era. In some cases, knowledge is lost altogether. Interestingly, the very act of translating tactic knowledge held by individuals into codified knowledge available to all can lead to a loss of knowledge.

Masking. One of the earliest features of genetic information of which scientists became aware was the ability of dominant information to "mask" recessive genes that may not make their appearance in the phenotype for several generations. The masking of recessive genetic information may be considered a form of falsification in the sense that the presence of certain information is hidden. The parallel in human communication would be misrepresentation through selective presentation of information.

Human Influences on the Facticity
of Genetic Information

Deliberately or not, people can falsify genetic information through manipulations of the genetic information (adulteration) or representations of it (misrepresentation). Because of the ancient link between political power and control over fundamental resources such as food, which are reliant on genetic information resources, unofficial genetic information has also often carried the connotation of unacceptable falsity.

Adulteration. Interactions among the handling of germplasm, natural differences between samples of biological material, and environmental degradation create multiple opportunities for adulteration. Blending opportunities arise at multiple points of the distribution process: on the farm, at the point of delivery to traders, during transport, in the course of consolidating small batches into larger ones, during milling or other processing, and in packaging (Thorbecke, 1992).

The difference between deliberate adulteration of grain and grain blending is one of degree. For example, it has long been accepted practice to combine grain from different truckloads of widely varying quality as a receiving practice during periods when high volumes of grain are being transported. For this reason, blending is both the stimulus to regulatory interventions and a factor that confounds regulatory success. The ease with which genetic material of differing qualities and characteristics can be blended and the dispersal of opportunities to do so is among the concerns on both sides of the debate over genetically modified (GM) foods discussed by Murdock (chap. 9, this volume) and Priest and Ten Eyck (chap. 7, this volume). Consumers claim they can never know whether they are getting GM foods, and the agriculture and food processing industries insist that, irrespective of labeling and any efforts at containment, no single entity can control what happens to the material it produces once it enters the distribution chain.

Overgeneralization and statistical aggregation offer equivalent opportunities for "adulteration" of the facts of human communication that can similarly be intentional or not. The opportunity for such adulteration of the facts has even been exploited in recent years by the statistical technique of "perturbing" individual detail systematically to protect the privacy of individuals reported upon in a sample ("Privacy," 1991). Adulteration in human communication also occurs when misinformation and information are combined.

Misrepresentation. As with the treatment of facticity in language, where rules of legal evidence and the illegality of libel, perjury, and fraud attempt to restrict deliberate falsification of information through misrepresentation, a great deal of law and regulation has developed in the effort to prevent falsification of genetic information through misrepresentation. As early as the 4th century BCE in Greece, special laws governed trade in grain in ways that went beyond laws applied to other types of commodities, and there were special magistrates to administer those laws—whereas misrepresentation in the course of trading in metals, textiles, or oil was punished by fines or imprisonment, misrepresentation of the genetic information of grain was punishable by death. In China, Confucian morality is said to have grown out of Confucius' experience in management of a public granary in his youth (Spitz, 1983). With the elongation of the marketing chain, opportunities for misrepresentation of genetic information by distributors and traders also multiply. As a result, numerous rules establishing standards for distinguishing among grades of grain, measurement, grain handling, and labeling have been put into place. In the mid-19th century, it was an offense in Britain to sell any "killed or dyed" seeds with intent to defraud, for example, and many U.S. states legally require seeds to be "true to name" (Kloppenburg, 1988). Misinformation, disinformation, deception, and fraud are the obvious equivalents in human communication.

Unofficial Genetic Information. Over the long course of human history, households have saved seeds from 1 year's crop to sow for the next even when those in power controlled the storage and distribution of grains. In flush times, political entities had little or no interest in control over such genetic information because it was for personal use only. In times of scarcity, however, governments were interested in collecting these hoards as a form of taxation that transferred control over genetic information from producers to nonproducers (Spitz, 1983). During the 19th century, however, governments became aggressively involved in collecting and distributing genetic material (both plant and animal) in the effort to improve the quality of what was produced in the private sector. At the same time, the processing, distribution, and marketing of materials such as seeds took on an industrial form (Duncan, 1989; Martinson & Campbell, 1980). As a result, by the late 19th century, farmers in many places could buy seed from either the government or private vendors. During this period government-distributed seed was considered to be the more valuable based on the perception that the government was less likely to misrepresent the genetic information and was motivated to distribute only high-quality seed. Government seed was cleaner, unlikely to be old, and free of grit and weed seeds. The government also kept raising its standards for seeds, instituting testing procedures for germination, cleanliness, and other features in 1886 (Kloppenburg, 1988). Ultimately this translated into an equation of unofficial genetic information with falsity (Tarrant, 1992; Thorbecke, 1992).

Here there is a parallel with libel law. The notion that truth is a defense was an innovation of the United States in colonial times; up until that point, and in some places still, merely disagreeing with official versions of the facts of the nature of government activity was or is treated as libelous. In such circumstances, government information is factual and all else is treated as false. Interestingly, just as struggles over the nature of facticity in reportage and public debate has been an organizing goal for civil society, so struggles for control over grain have been significant in the development of civil society historically (Benson & Shaw, 1992). Contention over the accuracy, validity, and utility of unofficial information lies at the basis of debates over the procedures of objective, *The New York Times*-style journalism and new journalism because the latter insists on utilizing and treating as important fact information that comes from unofficial sources.

GENETIC INFORMATION AND SECOND ORDER FACTICITY ISSUES

Second order facticity issues appear when genetic information is inserted into institutional frameworks such as those of the Patent Office, genebanks (genetic information archives), and labeling requirements. Such issues are deeply intertwined with those of standard-setting (Hill, 1990).

Patent Office

French rose breeders led a lobby for patent protections just like those of inventors of machines in the late 19th century (Mooney, 1988), but moral, conceptual, and logistical barriers stood in their way. Resistance began to fall in 1922, when Germany accepted a process patent on a bacterium, and a meeting of plant lawyers in London began to explore the possibility of protection for plant varieties (Crucible Group, 1994). This possibility became reality in the 1930s, when the U.S. Plant Patent Act allowed for the assertion of intellectual property rights in asexually reproducing plants and the Paris Union for global patent rights was amended to include flowers under patentable material. From that point on, one type of genetic information after another became subject to patent, with Germany again being the first to permit process patents for the breeding of animals. The U.S. Plant Variety Protection Act of 1970 made it possible to own patents on sexually reproducing plants, and in 1980 the U.S. Supreme Court accepted the patenting of microorganisms in the landmark case *Diamond v Chakrabarty*. In 1987, the U.S. Patent Office began to consider patents on animals, and in 1992, the first patent over an entire species was granted in the United States—a practice Boyle (1996) equated with granting Ford the patent on the car. In 1993, the U.S. government applied for patent rights over human cell lines of the citizens of several countries in the developing world, and harmonization of patent laws across nation-states became a requirement for participating in the global trading system.

The U.S. Patent and Trademark Office (PTO) has justified its changing position over time on the patentability of genetically engineered plants, seeds, and tissue culture as responses to the growing descriptive ability of molecular biologists and geneticists. A patent office "certifies" the facticity of genetic information when it assigns property rights upon demonstration of its uniqueness, accomplished through narrative. The raw materials with which biotechnology works are usually so complex and highly integrated before human intervention that it is not possible to describe constituent elements precisely. Thus, developments for which patents are sought are more likely to be described in functional or informational than in structural terms. All that is required is convincing a patent office that information sup-plied by the inventor sufficiently alters the nature of the genotype to be dis-tinguishable as something new (Bent et al., 1987). Replacing descriptions of specific exemplars with population parameters, as discussed earlier, is one approach for doing so. Providing descriptions that are loose or vague—of-ten making it possible for one genetic invention to receive several different patents—is another. With the growth in knowledge about genetic informa-tion, descriptive ability has risen, accelerating in turn the commodification of genetic information and an increase in the incentives to continue to de-velop such knowledge.

Some of what is presented to the Patent Office as unique, however, is not. Unlike the standard utility patent statute, beginning with the Plant Pat-ent Act of 1930, utility was not required to patent genetic information, only novelty and uniqueness. The question of relative superiority to existing va-rieties was also considered irrelevant. Thus, the number of named species protected by patent has proliferated even though these species often bring nothing new to either economic utility or crop performance at any stage of production, processing, or distribution. The result is that a high percentage of patents go to pseudo (false) varieties. In the mid-1980s, 62% of varieties protected by the Plant Variety Protection Office were accounted for by five crops only, almost all pseudovarieties. The Federal Seed Act of 1939 provi-sion making it illegal to use synonyms for a single variety has not prevented this type of falsification: It only requires that patent applications include a demonstration that some research had been undertaken to develop the item for which a patent is sought.

Although seed company executives claim that farmers cannot tell that competing brands are virtually identical, this is not the case. A 1980 study demonstrated that Illinois farmers were well aware they were being forced to choose among 253 different selections of the same species of corn (Kloppenburg, 1988). Serious conflict has thus arisen between farmers and the seed industry because of the rising costs associated with creation of pseudovarieties. Those in agriculture have sought certification programs separate from those of the Patent Office to cut through falsification of ge-netic information descriptions, but the seed industry opposes the move.

Intellectual property rights do not play the same role in certification of facticity that patents do for genetic information because of course copy-right does not require facticity. The one intellectual property rights issue that applies to both is patent "stacking" (English Nature, 2002). In the digital environment, sensitivity to the use of material owned by one entity in the production of further content has gone up. Stacking—linking several patents together to produce one item—has not yet been proposed as a way to cope with this in the copyright domain. In this period of extension and experi-mentation with intellectual property rights, however, that would be an interesting addition to laws dealing with compilations, collections, and derivative works. The multiplicity of patents involved in information technologies is a stacking problem that has long plagued those who try to enforce antitrust (competition) law.

Genebanks

The first order problem of decay becomes the second order problem of conservation when genetic information is archived in genebanks. *In situ* genebanks—"crop reservations" (Frankel, 1988)—preserve environments within which particular genetic information thrives so that it can be maintained on site in its natural context. Most genebanks are *ex situ*, off site, in scientific laboratories and botanical gardens.

The problem of decay requires high standards of quality control in genebanks for maintenance of their credibility. As far as possible, the genetic integrity of each individual accession is maintained, meaning that environmental influences on phenotypes must be controlled (Konopka & Hanson, 1985). Bibliographic data that justify inclusion of particular samples in the collection vary from genebank to genebank. They can include detail on breeding landmarks and special genetic characteristics, and always include data regarding the viability of seeds and regeneration and distribution instructions. Setting the standards for each type of data is a process of continuous negotiation between breeders and curators. It is the responsibility of curators to determine and standardize the threshold values for the number of seeds per sample in store and the regeneration standards that must be used to ensure that seeds remain viable, bearing in mind breeding systems, patterns of variability, and individual seed yield. Genetic information from different regeneration cycles is never mixed; information from different accessions of the same species are kept as separate samples, often in different locations to protect the distinction between them. Inaccurately classifying genetic information, a lack of fit between archival categories and empirical realities, differences in terminology from genebank to genebank, and active misrepresentation of genetic information to ensure its inclusion in genebanks are all potential sources of falsification during the process of translating genetic information into archival form.

Archiving, too, builds knowledge architectures when it effectively distinguishes between what will be accepted as fact and what will not. When the Chief Archivist of the United States (head of the National Archives and Records Administration) decides which U.S. government records are worth keeping for historical purposes, he or she determines the official view of U.S. history. George W. Bush has recently drawn dramatic attention to the political importance of this ability of archives to build knowledge architectures with an executive order making it much easier for presidents to forbid access to their papers as part of the mélange of policies putting in place changes in access to and use of information since the events of 9/11. In a related way, indexing fulfills a socially structural function via its determinations of knowledge architecture.

Most of the second order facticity issues in the world of social information deal with knowledge architectures. Some of these are textual: Compiling facts in dictionaries and encyclopedias makes them both official and available. Disputes over what should and should not be included provide interesting insights into social divisions. Eric Partridge, the lexicographer who seminally built dictionaries of slang, spent a life battling for acceptance of study of the language of those at the bottom, on the periphery, and in resistance.

Knowledge architectures are also built conceptually through the design of frameworks that are then used to guide institutional, economic, and legal behaviors. The classification system for industries and products used for purposes of economic analysis provides an interesting example; the long-standing Standard Industrial Classification (SIC) codes ultimately strayed so far from empirical realities that they are no longer in use. Although in 1997 a new conceptual framework, the North American Industrial Classification System (NAICS), came into use, analysts are still trying to figure out just how to make distinctions among industries and products within the information sector because, as they say, they can't quite figure out just what is being bought and what is being sold.

Institutions also have a knowledge architecture function because the very proceduralization of fact production within institutions has this effect. Analysts of journalism have discussed this in their analysis of the ways in which events become reportable facts when they pass internal institutional barriers—a license gets granted, for example, or someone graduates. This linkage of institutions with knowledge further supports the value of official versus unofficial fact.

Labels

Labeling has been an issue in the marketing of germplasm for a long time. One of the first pieces of business when the American Seed Trade Association (ASTA) was formed in 1883 was to agree to print disclaimers of performance on seed packages and a mutual acknowledgment that it was easy to market the same seeds under different names.

Among the several biotechnology issues currently the subject of intense debate is the labeling of those foods that include plants or animals that are genetically modified organisms (GMOs). There are "how to" and "whether or not" questions, both involving facticity. Standard-setting organizations (SSOs)—often the same groups that certify industrial technologies (Clark & Juma, 1991)—require adherence to specific types of descriptive terms. Although labeling developed over time as a cultural practice, today it is explicitly used as a form of information policy (Magat & Viscusi, 1992). Ratings systems and software filters are labeling systems applied to human communication.

Whether or Not. The National Food Processors' Association claims that two thirds of what is available in grocery stores has been modified. Along with numerous other groups, it is thus pursuing labeling as a way to warn people of potential health hazards. At minimum, selective and voluntary labeling is the best and most efficient way to diffuse perceptions of imposed risk (Douthitt, 1995). Almost all of the food with GMOs being produced comes from the United States, Canada, and Argentina, however, where labeling is not required.

Undeniably, meeting proposed EU standards or those being suggested for the United States would be logistically difficult because of the practice of mixing agricultural products from different fields and farms. Labeling also adds significant costs to processing and distribution and, it is claimed, reduces scale efficiencies (Miller & Huttner, 1995). Critics also maintain the law tries to draw distinctions between foods that are not chemically distinct. Even advocates of labeling are not clear on where to draw the line: Should a pizza be labeled if it contains cheese made with biotechnology produced rennet (60%–80% of current cheese; Thompson, 1997)? Chickens raised on feed from biotechnology engineered corn? Food from plants that are genetically engineered, but have no new genes or gene products in the edible parts?

We are beginning to see evidence of health effects in GM foods. As such evidence grows, so of course will public concern. The issue is currently on hold in light of world affairs, but it could easily trigger a trade war between the United States and Europe.

"How To." It would seem that the use of labels would reflect a commitment to facticity, but Wilkinson (1987) points out that labels instead often conceal rather than reveal essential facts because they fragment whole foods into constituent ingredients that often hold little or no meaning for consumers. Doing so reinforces the image of the food processing industry as the supplier of nutrients as opposed to food—a redefinition that in turn opens up new markets for biotechnology products in the form of "nonconventional" foods or elements of balanced diets. Although eating food is sensual and laden with cultural practice and meaning, deciding on a diet is conceptual and ridden with rules. This definitional shift is of importance from the perspective of facticity because it substitutes one set of referents for another.

The FDA's approach is that a food derived from a new plant variety must be labeled as such if it differs from its traditional counterpart to the degree that the common or usual name no longer applies to the new food, or if a safety or usage issue exists to which consumers must be alerted. Labeling must be both accurate and material. Because there is no evidence that bio-technologically modified foods systematically or significantly differ from other foods in ways related to either nomenclature or safety, there is no need to label, although there would be a reason to do so if the nutritional content changed (Miller & Huttner, 1995). The EU has approved rules requiring the labeling of any foods containing .9% material derived from genetically engineered organisms (GMOs) not previously approved, although they must be approved by the European Parliament and the 15 member states before they go into effect. These rules are extremely complicated—if the corn oil used in a particular mayonnaise is even 1% derived from GMO corn, it gets a label, but if an egg is raised by a hen on nothing but GM feedcorn, it does not get a label. Alternatively, a "no biotechnology" type label would protect those foods that had been monitored throughout the production and monitoring chain, but would leave most foods unlabeled (Thompson, 1997).

GENETIC INFORMATION AND THIRD ORDER FACTICITY ISSUES

Although first order issues derive from efforts to empirically capture genetic information, and second order issues from the institutionalization of genetic information, third order issues appear in the course of reportage and persuasion regarding either the viability of biotechnology for investment purposes or its health and environmental risks. As with digital information technologies, there are both utopian and dystopian views. Third order distortions of the facticity of genetic information are rife because of media willingness to follow often wild speculation in both directions. The politicization of biotechnology has taken place with little knowledge about actual perceptions of its risks (Douthitt, 1995; Siddhanti, 1991). The construction of narrative texts involves facticity problems that are cumulative, including those of the first, second, and third orders.

Investment

Although Rifkin's (1998) book describes the 21st century as the era of biotechnology, the same was said at the beginning of the 20th when it was believed that the new technologies would bring an end to hunger. German use of biotechnology to produce cattle fodder during World War I led to dreams that people ultimately would be able to convert the evening newspaper into sugar so rapidly that the protein produced could be consumed the next morning at breakfast (Bud, 1993). The most recent "geneticization" of the public mind via media campaigns in the 1970s generated investment confidence and public support for the then-blossoming industry. Genetics was once again seen as the next big wave of technological progress, and by the close of the decade the media were a captive audience for biotechnology. Each new scientific advance became a media event designed to capture investment confidence and public support, and market expectations and social benefits were often overstated (Krimsky, 1991).

A number of forces contributed to the nature of media coverage. There were economic pressures from advertisers because the food industry has provided almost 20% of advertising dollars for the media since the 1960s (Mamiya, 1992), and pharmaceutical advertising has become increasingly important. There was political pressure because trade in agricultural products was significant for the U.S. budget (Mayer, 1986). Information subsidies from trade groups, the organization of newsbeats, and a lack of scientific training on the part of most scientists (Kitzinger & Reilly, 1997; Wiegele, 1991) make it easier for news media to be more pro-biotechnology than critical. Meanwhile the public interest aspects of news about the food, agriculture, chemical, and pharmaceutical industries were more difficult to discern and typically receive relatively little coverage (Oliveira, 1992).

Risk

The media do of course pick up on stories with high drama, such as claims that because of biotechnology, the ground will freeze, harmful bacteria will run rampant, plant and animal species will mutate beyond recognition, and cancer will spread. Popular awareness of risks from biotechnology was stimulated by two novels—Michael Crichton's (1969) *Andromeda Strain* and Kurt Vonnegut's (1963) *Cat's Cradle*. Scientific concern over biotechnology had a prehistory in Vietnam-era concerns over the military purposes to which research and development (R&D) in general were being devoted. Biologists, chemists, and physicians were in particular disturbed about chemical and biological warfare and the use of Green Beret medical teams for political purposes in Vietnam. It was clear that technologies developed for counterinsurgency abroad had immediate domestic implications as well (Krimsky, 1982). In the 1970s, a group of scientists frustrated by the National Institute of Health's (NIH's) reluctance to closely examine release of genetically modified organisms into the environment began to publicize the issue, letting the public know that scientists were polarized regarding the safety of biotechnology. By the late 1970s, national environmental groups such as Friends of the

Earth, the Environmental Defense Fund, and the Sierra Club were involved in litigation over risk assessment procedures and related matters.

High drama was introduced into the debate by Jeremy Rifkin's use of a variety of media techniques, including confrontation. In 1977, he led a group of supporters in an invasion of a meeting of the National Academy of Sciences devoted to rDNA science and policy and disrupted it with guerrilla theater. Other tactics followed, generating media coverage that began to be successful as first Cambridge, Massachusetts, and then dozens of other states and municipalities passed laws prohibiting rDNA experimentation when it requires physical or biological containment. Congress was ultimately moved to hold hearings (Krimsky, 1991). There was a similar reaction to genetic engineering in Europe by the mid-1980s, and even more concern on that continent than in the United States over genetically modified (GM) foods by the mid-1990s (Gottweis, 1995). Rifkin's books (1984, 1998) have been critiqued for being scientifically unsound, but he continues to stimulate public debate about important issues.

Meanwhile biotechnology risk has become increasingly politicized. Experimental release of genetically engineered organisms (GEOs) is taking place in developing countries, sometimes unknowingly and without approval ("Bugged," 1994), and sometimes, as has been the case with storage of hazardous waste, a deliberate choice on the part of impoverished and marginalized communities to earn cash by doing so. The United States has accused the EU of increasing perceptions of risk by the very slowness of its legal response. Research on GM crops is now demonstrating risk beyond those of unintended release, such as an increase in the demand of such crops for water, freak and/or non-bearing crops, the danger that super-resistant bugs or weeds will develop, and greater sensitivity to UV-B rays (Benson & Shaw, 1992; Ehrlich et al., 1993). Some interbreeding with other plants has been found (Lean, 2003) as well as real environmental costs when insects or animals reliant on a particular food source cannot tolerate genetically modified plants—the Monarch butterfly was the first casualty of this kind. Interestingly, although some foods such as the bioengineered tomato have caused a lot of controversy, other transgenic foods have not. The medical industry accounts for 90% of the products of genetic engineering but it is food that generates about 90% of the controversy (Thompson, 1997).

ORDERS OF FACTICITY IN HUMAN COMMUNICATION

The distinctions among first, second, and third order types of facticity brought to our attention by the study of genetic information can enrich understanding of facticity in human communication in two ways. It provides a way to conceptually bundle issues that have been addressed in different literatures so their similarities and relationships can be seen: First order issues are raised by research methods, second order issues by knowledge architectures, and third order issues by narrative production. It also provides a way to conceptually unbundle issues that have been grouped together in discussions of facticity so that the unique characteristics of each—and their relationships with each other—may be more clearly seen.

THE OWNERSHIP
OF INFORMATION

5

Justifying Enclosure? Intellectual Property and Meta-Technologies[1]

Christopher May
University of the West of England

As we enter the new millennium, the use of powerful information and communications technologies in the biosciences has led to an expansion of the possibilities for the commodification of genetic information. This is hardly unproblematic. Intellectual property is grounded on an asserted metaphorical link between material things and ideas. As David Ritchie discusses at length (chap. 2, this volume), metaphors drawing a parallel between information (or ideas) and physical materials can be, at the very least, misleading. When intellectual property (re)constructs valuable knowledge as property, and critics argue this encloses (makes property of) what was previously part of the global knowledge commons, the area of dispute is the use of metaphor to drive a legal regime. In the case of the human genome, the actual idea of owning natural genetic resources is often highlighted, not least of all as the ownership of rights to human genetic information is often equated with the ownership of humans. I start by looking at some of the theoretical issues underlying these debates and then relate these problems to biotechnology as a site of meta-technological convergence (Braman, 1990/2002) before briefly exploring a possible political response. Indeed the interpenetration of information and biotechnologies as meta-technologies has made the question of markets for property in knowledge and information ever more crucial and in need of a political rather than a merely technical analysis.

A central issue in the history of intellectual property has been how to balance the private rights awarded to those who develop important (and socially valuable) knowledge, with the public good of freely accessible knowledge resources. The knowledge commons, from which intellectual property has temporarily rendered certain items as property, have always been implicitly recognized in law.[2] Commons are collectively owned (for knowledge and information at the global level we might say owned by humankind), which allows an immediate recognition of the costs imposed when knowledge passes from potential collective ownership to actual private ownership (even if this is only temporary). Indeed by making such property temporary (by limiting the duration of protection), socially useful knowledge is subsequently returned to these commons. This also recognizes that many aspects of new knowledge are actually drawn from the extant pool of information and knowledge represented by these commons. However, although such extraction from the commons was originally (in the 17th and 18th centuries) regarded as a privilege accorded only in certain circumstances, the subsequent history of intellectual property has seen it gain the status of a right.

Property is hardly "natural": As Walter Hamilton stressed, it has always been "incorrect to say that the judiciary protected property; rather they called that property to which they accorded protection" (quoted in Cribbet, 1986, p. 4). Property in a legal sense can only be what the law says it is—it does not exist waiting to be recognized as such, but rather is the codification of particular social relations, those between owner and nonowner, reproduced as rights. Karl Polanyi (1944/1957) suggests that the idea that labor, land, and money might be commodities required a "commodity fiction" to be developed during the transformation from feudalism to capitalism. The rendering of things not originally produced for sale as commodities required a story to be told about these resources, which was not necessarily linked to their real existence or production, but rather narrated a propensity to be organized through markets. To have a market in knowledge, this commodity fiction is again crucial and has made the protection of intellectual property rights (IPRs) an issue central to the political economy of the new meta-technologies.

This fictionalization is central to the Trade Related Aspects of Intellectual Property Rights (TRIPs) agreement, which is overseen by the World Trade Organization (WTO). I have discussed the negotiations and general implications of the agreement at some length elsewhere (May, 2000) and here merely note that encapsulated in this agreement is a forceful invocation of the knowledge as property argument. The TRIPs agreement represents the most developed and powerful instrument to construct a metaphorical link between property in knowledge and the legal mechanisms that have been widely developed to protect material property rights. It is the first agreement to be truly global in scope (with robust enforcement mechanisms through the WTO) and the first time all forms of IPRs have been subject to the same set of legal mechanisms. The agreement is "a remarkable symbolic document," which promotes a specific view of property and market relations as part of a (neo-) liberal agenda of global governance (Burch, 1995). Encapsulated in the agreement, originally drafted by lawyers representing a group of 12 U.S. multinationals (May, 2000; Sell, 1998), is an overwhelmingly Anglo-Saxon legal discourse presenting a specific view of the justification of IPRs and the efficiency benefits from making knowledge and information property.

Much of this legal discussion of intellectual property assumes there is a clear metaphorical continuity between property in things and property in knowledge. Yet the institution of property in knowledge constructs a scarcity where none necessarily exists. This scarcity needs to be constructed because knowledge, unlike physical property, is not rivalrous. Where things exist in time and space, which is to say they are material (and may be subject to property rights), they cannot be used in two different locations at the same time or in the same location without some loss of utility to co-users. If I sell you my hammer, I cannot subsequently use it; if you and I both want to use the hammer (even if I agree to you using it at the same time), the ease of use for each of us is reduced by the need to coordinate our hammering. Information and knowledge are not generally like this.

Certain sorts of information may be rivalrous, especially where such knowledge or information may allow the owner to produce material goods for sale or take advantage of market conditions (differences in prices between markets, for instance). Inasmuch as knowledge produces material goods or market advantage, it may be rivalrous in the sense that non-rivalrous use would mitigate advantage and thus reduce the recoverable price from knowledge use. Thus, in a capitalist economy, this construction of rivalrousness becomes the central role of intellectual property. Although contemporaneous usage detracts little from overall social utility, without imposed rivalrousness, the ability to profit from the use or sale of socially useful information or knowledge would be constrained. Recognizing the constriction of social utility presented by intellectual property, other reasons need to be used to justify a regime of property in knowledge, and it is to those justificatory schemata I now turn.

THE JUSTIFICATION OF INTELLECTUAL PROPERTY AND THE CONSTRUCTION OF SCARCITY

Conventionally there are two philosophical approaches to justifying property and one more pragmatic justification, all of which are used to legitimize and support intellectual property in varying combinations (May, 1998,

2000). Not only commentators, but also legal documents and judgments, sometimes explicitly, but more often implicitly, draw on these material property-related narratives to justify the recognition of property in knowledge. Most important, the justifications are used in the TRIPs agreement and have been mobilized in the cases brought to the WTO's dispute settlement mechanism. Therefore, they play a profound and important role in the way the global regime of protection of IPRs is governed (and developed through legal precedent).

The first schema argues for labor's desert: The effort that is put into the improvement of nature requires it should be rewarded. In John Locke's influential formulation, this was modeled on the improvement of land. The application of effort to produce crops and/or improved resource yields justified the ownership of specific tracts of land by whoever worked to produce such improvement. Starting from this initial position, Locke then argued there was also a right in disposal mediated by money. This led him to conclude that all property, even after its initial sale or transfer, could be justified on the basis it had originally been produced through the labor of an individual. More important, property was also justified because it encouraged the improvement of nature through the reward of effort. Therefore, the Lockean argument supports property by suggesting that property encourages individual effort through the reward of ownership of the fruits of work. In contemporary debates around intellectual property, the argument that patents and other intellectual properties reward the effort that has been put into their development (the research investment made to develop a patented innovation; the marketing expense in establishing a trademark) is commonplace.

However, sometimes this argument is supported through the mobilization of a secondary justificatory schema—the notion of property's links with the self as originally proposed by Georg Hegel. Here the control and ownership of property is a significant part of the (re)production of selfhood inasmuch as selfhood relates to the establishment of individual social existence. It is the manner in which individuals protect themselves from the invasions and attacks of others. For Hegel, the state legislates for property as part of its bargain with civil society. Individuals allow the state to operate in certain areas, but protect their individuality (and sovereignty) through the limitations that property rights put on the state vis-à-vis the individual's own life and possessions. In intellectual property law on the European continent, this supports the inalienable moral rights that creators retain over their copyrights even after their formal transfer to new owners. In Anglo-Saxon law, this mode of justification has been less well received due to its implications for the final alienability of intellectual property. Nonetheless, especially where "passing off" of trademarks (the unauthorized use of logos and brand names often on substandard goods), and the pirating of copyrighted material (e.g., sampling of music) are concerned, this justificatory schema can sometimes be noted in the calls for redress based on the diminution of reputation or the ownership of (self-) expression.

There is a third set of additional and important justifications that often underpin the role of intellectual property in the realm of the meta-technologies. In this pragmatic or economic argument, the emergence of property rights is presented as a response to the needs of individuals wishing to allocate resources among themselves (May, 2000). Thus, North (1990) argues the enjoyment of benefits (and the assumption of costs) takes place in social relations through the mobilization of useful resources. The institution of property arose to ensure that such resources have attached to them the benefits (and costs) that accrue to their use, and this increases efficiency. In this story, property rights took the place of social (trust) relations and allowed complex trade relations to form over distance. Part of the continuing fluidity in the legal constitution of property rights has been the widespread attempt by owners to secure benefits while keeping costs externalized. Social efficiency would be best served by costs accruing to the property that delivers the benefit. However, for individual owners, it is more efficient to have the costs met by others.

Mobilizing a history of material property, this third schema suggests that efficient resource use is established through the use of markets in which property is exchanged and transferred to those who can make best use of it. Therefore, the development of modern economies is predicated on the institution of property and its ability to ensure the efficient use of limited resources. In this justification, it is this efficiency requirement that drives the historical development of property rights and now underpins the commodification of knowledge. This (institutionalist) retelling of the history of property carries with it the notion that property arose to ensure the efficient allocation of scarce economic resources. Even when it is accepted that this allocation may not be optimal, property rights are still presented as the most efficient method of allocation available, although they often produce a less than perfect solution. In the interests of efficiency, property as an institution is reproduced (and improved) through its legal and social use. This narrative of the efficient allocation of scarce resources is then brought to bear on the allocation and use of knowledge by meta-technological industries in contemporary society.

As a subset of this third justification (but linked to the first), one of the most common arguments utilized to substantiate IPRs is the need to support innovation. Drawing from Locke the notion of reward for effort in improvement, and from the third schema the idea of social efficiency, it is often asserted that without IPRs there would be little stimulus for innovation. Why would anyone work toward a new invention, a new solution to a problem, if they were unable to profit from its social deployment? Thus, not only does intellectual property reward intellectual effort, but it actually stimulates activities that have a social value, and therefore serves to support the social good of progress. Underlying this argument is a clear perception of what drives human endeavor—individual benefit. Only by encouraging and rewarding the individual creator or inventor (with property, and therefore market-related benefits) can any society ensure that it will continue to develop important and socially valuable innovations, which serve to make society as a whole more efficient.

All this indicates that one of the central purposes of IPRs is to construct a scarcity (or rivalrousness) that allows a price to be taken and knowledge to be exchanged in market mechanisms to further social efficiency. In a clear statement of this requirement (utilizing the labor desert schema), Arrow (1996) noted that if

> information is not property, the incentives to create it will be lacking. Patents and copyrights are social innovations designed to create artificial scarcities where none exist naturally.... These scarcities are intended to create the needed incentives for acquiring information. (p. 125)

The construction of scarcity through the commodification of knowledge plays a vital role in the operation of modern capitalism. Yet considerable effort is required to support the argument that such scarcity is socially beneficial, and in one way or another such claims frequently draw their inspiration from an assertion of the "tragedy of the commons." Although originally developed as an account of historic overuse and degradation of environmental resources, this position is frequently either explicitly (e.g., Carr, 2000) or implicitly appealed to regarding knowledge in economic relations.

SCARCITY AND THE TRAGEDY OF THE COMMONS

Hardin's (1968/1993) famous rebuttal of the social efficiency of the commons is formulated in the first instance as a story of grazing livestock on common land. While the land's capacity to support the herds is above the total number allowed to graze by all the herdsmen, life continues happily. However, an increase (through a decline in mortality, better health, or less poaching and disease) up to the maximum capacity that can be supported by the common land changes things. Each herdsman can now only benefit from adding an animal to the herd at the cost of overgrazing the commons. The benefit of adding an animal is fully captured by the herdsman (either by the subsequent sale or continuing use of the animal), whereas the cost represented by the problem of overgrazing (a decline in food available to each animal and the eventual degradation of the resource) is shared equally by all the herdsmen. Acting rationally, each herdsman will note the costs of adding an extra animal (which are shared) are far outweighed by the benefits captured for himself. Each acting in their own self-interest maximizes their herd, degrading the commonly held resource and resulting in the final destruction of the common land through overgrazing. As Hardin (1968/1993) put it: "Freedom in a commons brings ruin to all" (p. 9).

The answer is not to forbid the use of the commons altogether because then no prospective benefits can be gained for society. Rather the question becomes: "How do we legislate temperance?" (Hardin, 1968/1993). Social arrangements must be changed to reduce the possibility of the tragedy developing, but for Hardin any "alternative to the commons need not be perfectly just to be preferable" (p. 12). This leads him to suggest that, despite its flaws, a property rights system alleviates the tragedy. Although it replaces merit or justice with wealth as the key allocation mechanism, "injustice is preferable to total ruin" (p. 17). Property rights, whatever their disbenefits and injustices, by virtue of their support for the overall public or social good of the avoidance of universal ruin are justified even when individual owners enclose previous commons. Criticisms of such new arrangements are likely to be short-lived:

> Every new enclosure of the commons involves the infringement of somebody's personal liberty. Infringements made in the distant past are accepted because no contemporary complains of a loss. It is the newly imposed infringements that we vigorously oppose; cries of "rights" and "freedom" fill the air. (Hardin, 1968/1993, p. 18)

If time brings acceptance of enclosure, then the technical solution of property rights can override any short-term objections because these critics fail to understand the bigger problem. Once the solution has been imposed and the political debates have exhausted themselves, behavior will adapt, and the tragedy will not only be averted, but will be forgotten with the rules becoming part of the generalized legal organization of society.

This suggests that arguments regarding the enclosure of knowledge may merely be transitory, are to be expected, and can be safely ignored. Once the generation that disputes the TRIPs agreement, or its application in the realm of biotechnology, has faded from active politics, these arrangements will become part of the normal political landscape. Once again this rests on the maintenance of the metaphorical link between material property and property in knowledge. Importing these arguments into the debates over intellectual property requires a particular view of what constitutes the social good in knowledge. Starting from the premise that labor should be rewarded with property and, more important, that no one would undertake meaningful labor without such reward, it is then proposed that it must be a social good to support and encourage innovation and intellectual endeavor, not least of all because of the advances that will then be supported. By imposing individual private property rights on the knowledge commons (where innovations might otherwise reside once developed), the tragedy of underinnovation is avoided. When knowledge is freely available, the stimulus to innovate is absent. Although private property rights may produce some injustices, they will increase the overall level of intellectual production and thus serve the social good. However, if the social good served by knowledge and information privileges availability, which is to say if the diffusion of knowledge is a social good, the enforced scarcity of intellectual property becomes problematic.

The process of enclosure explicitly downplays the possibility that knowledge can be used without depleting its intrinsic value to society as a whole: IPRs are defined against the notion of (economically) freely available knowledge. This is tempered by the formally limited duration of protection for IPRs, which allows the final return of information and knowledge to the commons. Although it may be possible to argue, as many critics of IPRs do, that to commodify knowledge "makes ideas artificially scarce and their use less frequent—and, from a social point of view, less valuable" (Vaver, 1990, p. 126), this claim is seldom accorded significant weight where meta-technologies have widened the possible scope of intellectual property. That said, there is an increasing move to publish scientific studies in noncopyrighted electronic journals to allow free access and a growing campaign for freely accessible "digital libraries," which some commentators feel will fatally undermine intellectual property. If knowledge can be used in a nonrivalrous way, if people can use the same knowledge at the same time without any impact on the benefits from the use of such knowledge, then the rendering of a scarcity in such knowledge (to establish a price) also inflicts a social cost on those to whom use is blocked or limited by price. Although meta-technologies have highlighted this concern, technical solutions (such as electronic publishing) are not the only approach.

Arguably the concern for a social realm of free knowledge is already encapsulated in the limited duration of IPRs; there is a balance between private benefit and social good in intellectual property's legal construction. The owner of intellectual property is awarded a monopoly over the particular knowledge object to allow a return or reward to be gained, but this is tempered by the need to allow full and free availability of this new (and useful) knowledge once the period of protection has expired. Thus, for patents (whose social availability is regarded as important), protection is currently set at a minimum of 20 years by TRIPs to allow for a reward to be earned before the subsequent open-ended period of social use. For copyrights (where social utility is seen as marginal), creators are protected for their lifetime and their descendants for a minimum of a further 50 years.[3] These periods of protection have been established over the history of intellectual property to balance the private rights of innovators and authors against the public rights of society as a whole. Does this balance as currently constructed unacceptably limit the extent of the contemporary global knowledge commons?

PROPERTY AND WITHHOLDING

For property the relationship between private and public benefits is the history of the move from the common understanding of property as physical things held for the owner's use to the more modern conception of property as assets, which can be used or otherwise sold to another potential user. However, when the recognition of property rights is co-existent with scarce resources, then,

> the mere holding of property becomes a power to withhold, far beyond that which either the laborer has over his labor or the investor has over his savings, and beyond anything known when this power was being perfected by the early common law or early business law. (Commons, 1924/1959, p. 53)

It is this move from holding to withholding—the ability to restrict use—that is of crucial importance. When the resources required for social existence are scarce, the distribution of the rights to their use (property rights) becomes a central, if not *the* central, issue of political economy. Conversely, when such resources are potentially freely available (as is knowledge), the imposition of property rights introduces this scarcity and supports the development of social power with regards to the benefit from their use.

However, the history of the recognition of property in knowledge has not been entirely patterned by the recognition of exclusive rights to withhold. Rather IPRs owe their origin in the 17th century to the grant of privilege to individuals in exceptional circumstances—occasionally for the introduction of a new technology, but more often for a particular service to the Crown or to reward a member of the Royal circle (Sell & May, 2001). By reinscribing the monopoly privileges accorded to the holders of intellectual property as rights, the common interest of all individuals in upholding IPRs as part of a social system of rights is asserted, and the possibility of obligations incumbent on the knowledge owner is obscured. The original bargain of intellectual property—that if privilege is being awarded, then some form of connected duty should also be accepted—is forgotten. Although patents originally emerged as grants of monopolies for other (often economic) reasons, they were regarded as privileges or distortions that were only justified on the basis of their other benefits (often delivered by the holder). The subsequent history of intellectual property has reversed this logic to make intellectual property the right that benefits society. In the last 200 years or so, this has led to a continual diminution of the possibility of a knowledge commons (a realm where information and/or knowledge is not subject to any form of private ownership).

The conception of intellectual property solidified for copyright with the rise of the romantic notions of individual creativity in the 18th century (Geller, 1994). With the need for patents to be lodged by a legally constituted individual, a similar norm has operated for patentable ideas since the earliest technical and industrial patent monopolies were awarded (Boyle, 1996). Intellectual property discourse increasingly drew on the notion of the individual creative individual, the *author*, acting in solitude to produce new knowledge. This eventually constructed an *empire of the author*, where all knowledge has a moment of genesis that justifies the IPRs attached to them (Aoki, 1996; May, 2000). This has spread from its original limited coverage (under both patent and copyright law), which sought to protect only innovative knowledge and the expression of individual's ideas, to more recent moves to widen the forms of knowledge and information covered. This broadening of protection now sometimes covers patterns of collected information, such as directories or reference works where courts have regarded the effort of collation as justifying the reward of protection. Compilers sometimes include a few false entries to enable them to demonstrate a rival has copied their directory rather than (re)compiled a new one, expending little protection-worthy effort. Protection has also been expanded to include the fruits of discovery, where the discovery is codified into a form that is original, genetic information being perhaps the most apposite example of natural, and thus formally public, information being rendered as private property. The global knowledge commons represent a major resource from which new knowledge can be developed and enclosed as private property.

The idea of the value of knowledge commons was originally the cornerstone of intellectual property. However, when examined closely, these commons have now become only residual, and the proliferation of private ownership rights is diminishing them (enclosing the knowledge and information the commons once contained). The public good of knowledge production has required the limitation of the knowledge commons. Now it represents only whatever is not claimed by the rights accorded to private intellectual property. There are therefore no effective safeguards to halt its erosion; the expansion of private rights in knowledge has brought about the knowledge commons' decline.

Conventionally when intellectual property is eulogized, it is on the basis of the protection of the creator—the owner of such knowledge that is made property. Their rights are protected so as to act as a general spur to innovation and socially useful activity. Arguments about just dessert and selfhood are allied to the need for social efficiency in the allocation of resources. However, as Waldron (1993) notes, all this talk of property "sounds a lot less pleasant if . . . we turn the matter around and say we are imposing *duties*, restricting *freedom* and inflicting *burdens* on certain individuals for the sake of the greater social good" (p. 862). That is, IPRs limit the actions of others regarding knowledge vis-à-vis the owners of intellectual property; as such these others are being forced to sacrifice their particular wants or needs on the alter of social necessity. Nonowners' rights are constrained because these rights are regarded as less important in law than the support of the social good of innovation by IPRs. As Waldron pointed out in the realm of copyright, the limitation on activities (unauthorized copying or plagiarism) is hardly life threatening. Yet what might be the reaction if we look at areas where intellectual property limits the use of knowledge, which has a direct impact on relative welfare, in the realm of the new meta-technologies, for instance? The real consequences of the distribution and control of intellectual property rights lead to a more critical conclusion regarding the social good served by their general protection.

The social costs of withholding information and knowledge may actually be quite different in scope and importance to that suggested by the prior narratives. The public good we should be concerned with may not be rewarding innovation, but rather the subsequent use of such innovations. Yet even arguments supporting some modification in the award of intellectual property rights often tend only to examine explicit economic costs rather than questions of social justice (see e.g., Dawson, 1998). Social justice is not necessarily absent, but rather is deferred. The emergence of meta-technologies has led many to question the justice of this deferment as its explicit social and welfare costs are now quite evident. As I already noted, knowledge is characteristically nonrivalrous; it is also formally nonexcludable; once an idea is known it cannot subsequently be forced back into the unknown. To ensure a continuing supply of innovation (the development of knowledge seen as a public good), states' laws have legislated a time-limited enclosure to encourage intellectual effort. It is to the social consequences of one particular enclosure that I now turn.

META-TECHNOLOGY, THE HUMAN GENOME, AND INTELLECTUAL PROPERTY

Scientists have now largely completed the construction of the picture of the human biological blueprint, the Human Genome Project, to much celebration and the expectation of wide social benefits. Utilizing powerful information technologies and cutting-edge biotechnological science, the genetic structure of our species (if not yet fully understood) has been revealed. However, all is not well:

> On stage at the world-renowned Pasteur Institute in Paris, Craig Venter, one of the world's foremost biological scientists, announced he was just weeks away from completing the greatest scientific endeavour that mankind has ever known: the human genome, the book of life of *Homo Sapiens*. It should have been a victory lap, applause, praise, talk of Nobel prizes and commendations from his peers.
>
> But instead there was silence. For most of the 600 academics in the audience, gathered from research institutes across the globe, Venter was the traitor; a ruthless competitor who sold his soul to American business and who now plans to charge the rest of the human race a fortune to read our own genetic code. (Toolis, 2000, p. 11)

As this report suggests, although this is undoubtedly an important scientific advance, it also represents a high-profile manifestation of a political issue that will become increasingly pronounced as meta-technologies have more and more impact on society: the ownership of socially useful information and knowledge.

This is not the first time Venter has tried to establish a monopoly over genetic information. In the late 1980s, Venter's researchers were constructing a library of expressed sequence tags (ESTs), which indicate where particular groups of human genes are within the entire sequenced genome. The rights to access this information were sold to SmithKline Beecham (SKB) and others through licensing agreements, which also enabled the owners and developers of the EST database, Human Genome Sciences (HGS), to demand fees from any subsequent commercial application developed by the licensees. Those companies unable or unwilling to pay were denied access to the data. However, Merck (one of SKB's rivals) set up Genbank as a public depository for such information discovered by other teams of researchers. This undermined the value of HGS's library because at that time only the database was legally protected, not the ESTs. Once a rival source of ESTs was available, this undermined the price HGS could charge for access to their information.

Merck's motivation was not disinterested philanthropy. The company saved itself considerable access fees that it might have otherwise had to pay: "Merck expect[ed] to profit more in absolute terms by making the database publicly available, even if other firms also profit[ed] as a result" (Eisenberg, 1996, p. 570). Recognizing that it was not the gene sequences that were most valuable, but rather the uses to which this information was put that would generate major profits, Merck banked on its comparative advantage in product development to capture the benefits of publicly available (raw) genetic data. However, in the following years, the biotechnological industry has sought to control this information as a way to also capture the benefits from any subsequent use found for the sequences.

Genetic patent applications are flooding into the U.S. Patent and Trademark Office (U.S. PTO), yet in the majority of cases the potential use remains unknown. Yet only by lodging patent applications for specific ESTs can companies attempt to ensure that when a product is developed (by them or by another company) they will be able to secure their reward. However, patent awards have historically been based on three criteria: newness, nonobviousness, and usefulness. Although these sequences may be new and the test of nonobviousness may be argued (nonobviousness is related to the impossibility of a practitioner adept in the field independently coming to the innovation in the normal practice of the art), usefulness seems indefensible. Whatever the other considerations, which are many, most genetic-material filings do not warrant patentability without an established utility. Many are actually submarine patents intended to remain hidden, but pending until some other patent reveals a use, when the submarine patentee can then show an earlier initial filing that invalidates the later patent (but enables the capture of the revealed use).

Thus, it seems many companies are patenting widely and hoping they will be able to gain some strategic advantage from the applications that can finally be defended (even if this is a minority of their claims; Thomas, 1999). With overworked and underpaid examiners at patent offices in Europe and America, who lack the requisite skills (or perhaps, more important, the time) to examine these applications fully, more often it is left to disputes to clear doubtful applications from the system. Furthermore, those lawyers who show talent in the area are soon tempted to take many of the well-paid private-sector consultant jobs on offer from companies preparing applications. Well-funded applicants stand a much better chance of defending claims than politically driven disputants. All this seems some way from the ends intellectual property is supposed to serve if we take its justificatory schemata seriously.

Essentially, the U.S. PTO has relaxed its criteria for awards while also expanding its recognition of patentable activities. (Indeed this is not the only field in which this has happened, with recent patent grants to one-click ordering and other information economy business practices, the Office is allowing an expansion of the types of knowledge resources that can be subject to protection.) A dynamic of relaxation and expansion has been evident since the Plant Patent Act of 1930 originally allowed the patenting of products of nature (reversing the previous explicit prohibition) provided they reproduced asexually. This was expanded to new and distinct sexually reproducing plant varieties by the U.S. Congress in 1970 (Fisher, 1999). A crucial relaxation finally stimulated widespread moves into patenting genetic materials: An appeal in 1980 to the Supreme Court regarding a 9-year-old dispute between Indian microbiologist Ananda Chakrabarty (claiming a patent on a genetically engineered microorganism that consumed oceanic oil spills) and the Patent Office, which had refused his original claim, was settled in favor of the grant of patent, ruling that whether an invention was animate or inanimate had no bearing on its patentability provided the three usual criteria of novelty, usefulness, and nonobviousness were met. The manipulation of nature could produce patentable products.

Despite the original hopes of the Court that this precedent would be narrowly construed, it actually became the basis on which patents for genetic sequences would be awarded in the subsequent years, and the U.S. PTO has "moved sharply in the direction of strong and broad patent protection in biotechnology" (Maskus, 2000, p. 55). However, the U.S. PTO also started to allow applications where genetic material had been discovered, and not necessarily engineered, if such material were a purified natural substance, allowing that a purified bacterial culture was the "product of a microbiologist" (Golden, 2001, p. 125). This suggested that two grounds for patent (newness and nonobviousness) might no longer have to be fully fulfilled for a grant to be deemed actionable. Although historically case law in the United States has differentiated between discovery and invention, this is not explicit in the underlying statutes (Sedjo, 1992). The difficulty of differentiating between discovery and invention when natural processes are being manipulated has now seemingly been settled in favor of characterizing it as invention, at least by the U.S. PTO if not by many critics of patenting in this area.

These settlements at the PTO have involved an implicit compromise of patenting criteria. Given that it is only the combinations of organisms that may be new, newness is hardly unproblematic for genetic inventions. Furthermore, because the manipulation processes have been largely standardized as part of the development of the biotechnology industry, the idea that any result would not be obvious to a skilled practitioner of the art (the criteria of nonobviousness) also seems strained at the very least. As Golden (2001) argued, especially with regard to genetic material, the utility function (that patent applications must define a specified use) has also been relaxed to allow computer-modeled evidence for proposed utility rather than actual demonstration of use, although *bioinformatics* is notoriously unreliable, with a failure rate as high as 95% by some accounts. Nevertheless, the U.S. PTO granted 2,330 patents covering gene sequences by the end of 1999, and this by no means represents the majority of filings (some reports suggest over half a million applications are pending).

The flood of filings, alongside the confusion about which are novel and which might duplicate each other, has prompted the PTO to rework its award criteria (Garber, 2000; Van Brunt, 2000). Thus, by no means all (or even the majority) of the applications are likely to be granted nor patents awarded upheld, not least of all it now seems likely there are only around 30,000 genes in the human genome. Nevertheless, the "gold rush" continues in hope that some of the claims will be defensible and thus vastly profitable. As Kahin (2001) notes in his discussion of how the U.S. PTO is always in danger of being captured by well-resourced applicants (examiners are essentially overworked and underpaid), the PTO defines its mission relative to patents in the following terms: "The primary mission of the Patent Business [sic] is to help customers get patents" (quoted in Kahin, 2001, pp. 9–11). Thus, given institutional pressure to "help customers get patents" and the disparity of resources between the PTO and applicants, it is hardly surprising the path of least resistance has been to issue patents and let appeals sort out the problem cases. Of course this means that, although high-profile cases may have the patents refused on appeal, there is the distinct possibility that doubtful cases will get through because no one has the resources or will to dispute the grant.

On the other side of the Atlantic, despite the European Patent Office's (EPO's) previous practice of blocking patent applications that might be regarded as morally suspect, the Office has never expressly excluded genetic material. Although the notion of *invention* in patent law on both sides of the Atlantic has always been quite permissively conceived, as meta-technologies have emerged the negative prescription of invention (i.e., ruling out what is explicitly *not* invention) has allowed the patenting of human genetic materials (or more accurately their ESTs) to proceed with little obstruction. Thus, under the European Patent Convention of 1973, such patenting is allowed because it is not expressly forbidden by the three limitations therein: methods for treatment by therapy, natural biological processes for producing plants or animals (and their results), and invention contrary to public morality.

According to Margaret Llewelyn of the Sheffield Institute for Biotechnology Law and Ethics, in these cases,

> Morality appears to be determined by a two-part test: whether the public would find the invention "abhorrent"[4] and even if it does, whether the invention has beneficial uses. If convinced of the latter, even if this is a speculative benefit, the patent office has proved reluctant to deny a patent on grounds of the former. (Llewelyn, 2000, p. 32)

Where it is assumed the health impact would be beneficial, it is regarded as self-evident that any moral concerns would be rendered immaterial: Given the choice between life and morality, the EPO assumes we will choose life. Thus, patent applications in this area now usually also include detailed speculation about the likely future health benefits of products drawn from the patented material. Like other areas of intellectual property law, where there is a choice to be made, the claims of the potential owner of intellectual property are favored (May, 2000). Thus, the enclosure of the available knowledge and information resources (in this case linked to the human genome) continues.

In this regard, the patenting of DNA has been judged by the EPO (in a judgment upholding the patent for H-2 Relaxin in 1995) to be the protection of specific information, not the patenting of life (Crespi, 2000). The genes remain in the putative commons, whereas the information about them and their use is protected. Although the commons remain (as formally they always have in patent law), given the techniques to utilize such information are remarkably standardized (not least of all as they were originally subject of a Stanford University patent) and expensive to complete, an effective toll (or rent) for use has been established. This is explicitly recognized in the "confirmation" of practice embodied in the EU's Biotechnology Patent Directive (98/44/EC) of 1998, which states:

> 5.2. An element isolated from the human body or otherwise produced by means of a technical process, including the sequence or partial sequence of a gene, may constitute a patentable invention, *even if the structure of that element is identical to that of a natural element.* (cited in Crespi, 2000, p. 177; italics added)

Once removed from the human body, any conception of commonality evaporates, and the identification of the sequence or partial sequence is enough to ensure a claim can be lodged. As in the United States, the historic criterion of patent novelty has been relaxed to ensure the protection of the results of the utilization of meta-technology in the biosciences. Furthermore, the emergent patenting strategy in the sector seems to be to "seek the broadest patent scope possible and to place considerable emphasis on the use of compromising language to maximise the reach-through claims of partial sequences" (Thomas, 1999, p. 138). Submarine patents are the norm, and the commons is shrinking.

SOME IMPLICATIONS

A decade ago, near the beginning of the mass mapping of the human genome, Kevles and Hood (1992) warned that the human genome was "in principle common property." It

> should be maintained as such as a matter of practical equity, since the mapping and sequencing of genomes will be—is already—the product of the inge-

nuity of a multinational community of scientists and of investments by many countries. Hard and imaginative thought needs to be given to means of preserving what is rightly common property while preserving incentives for private development of research for human benefit. (p. 315)

Unfortunately, the persuasiveness of the knowledge as property doctrine, encapsulated in intellectual property law generally and more recently specifically in the TRIPs agreement, has produced a particular settlement that does little to protect or value the knowledge commons, nor extend the commons to cover the human genome. Hard and imaginative thought has been constrained by the dominant view of intellectual property and the power of its justificatory schemata.

From the perspective of scientific ethics, one of the most often voiced objections to the patenting conducted by companies involved in genetic research has been the costs such practices impose on other experimenters, as well as the limitation to the flow of scientific knowledge, which might otherwise be freely available to the scientific community (Crespi, 2000). Although in one sense patents are an improvement over trade secrets, because patents must be made public to garner protection (by being lodged in an application at the Patent Office), and over copyright, where use can be restricted even when a payment is proffered, patents remain restrictive. Where once information might be freely used, now the awareness of possible patent infringement and the fees that will need to be paid for the use of patented knowledge resources compromises (or at least makes more expensive) the free run of scientific experimentation. In this sense, the commercialization of scientific endeavor is limiting the general benefits that could be derived from open communication of results.[5] However, the logic of patenting requires that once they are allowed, for any researchers in the field the lost reward of not patenting may increase along with the costs of licensing.

Whatever scientists may feel about the overall morality of limiting access to their results, in the main if others are patenting, then to not do so oneself is to assume an unnecessary cost disadvantage: paying for access but receiving nothing for the fruits of one's own endeavors when used by others (Merges, 1996). Thus, for American publicly funded research, the Bayh–Dole Act of 1980 was of profound importance. By giving universities the right to retain title to and license inventions that stemmed from federally funded research, research-based institutions could (and did) develop and maintain strong patenting strategies (Maskus, 2000). As a result, in the United States around 40% of human DNA-related patent applications have been filed by public sector researchers (Thomas, 1999). Although not as developed, a similar tendency is becoming apparent in Europe. As the trade applicability of "pure" bioscience has become more obvious, so more science has become trade related and thus open to the pressures to be made property.

Quite apart from specialist access, there is also a question regarding the justice of patenting aspects of a resource that might be (arguably) regarded as the property of all humans. If we all share the vast majority of genetic code (as the Human Genome Project has underlined), should companies be allowed to own particular parts of sequenced genes? These issues have been at the forefront of much of the popular disquiet regarding patenting of human (and other biological and agricultural) genetic information (see e.g., *The Guardian*, 2000). However, as one might expect, the biotechnology companies involved have repeatedly invoked the labor desert justification for patenting: Without such a reward, how would they earn a return on the investment? Without the promise of a return, how would they secure further funding to continue research? Any moral problems have been regarded, as Llewelyn suggested, as secondary to the interests of discovering new uses for genetic profiling and information and the reward for such work. Yet many of the promised benefits of this genetic revolution have actually been slow to arrive (indeed many have proved, at least currently, more chimerical than actual).

If ownership is to be asserted on the basis of the financial reward for effort, the actual history of biotechnology deserves a little attention. The early history of biotechnology is not a story of private sector enterprise, but rather the continuing and long-term support of scientific endeavor by various states' governments. Yet Eisenberg (1996) notes:

> The patenting of ESTs and the effective limitation on their use constricts their immediate use by other (competing) users of meta-technologies. This may have social costs, not least of all as swiftness to patent may not necessarily indicate that the most socially beneficial user or developer owns any particular sequence. (p. 572)

However, because the European Patent Convention clearly states that "The Task of the European Patent Office shall be to grant European Patents" (cited in Leese, 1996, p. 188), we should not be surprised that, like the U.S. PTO, it errs on the side of granting patents. As Leese (1996) points out:

> The wording may merely be the result of an enthusiasm on the part of legislators about the benefits of patents, or simply a formality of expression, but it hardly sets the tone for symmetrical assessment. Whether or not patenting is an overall benefit, in social and economic terms, its effects are certainly significant. (p. 189)

The main effect of this mission is the systematic preference for the private sphere of property rights over the public sphere of the knowledge commons, however such research has been funded. It is not clear that the use of this information will automatically bring benefits.

The ability to identify who is at risk from particular genetically derived health conditions, for instance, may be a mixed blessing: In a privatized health insurance system, even before they have developed symptoms individuals will see their health insurance costs spiral. Companies will no longer be making a risk analysis, but rather will bill on the expectation of illness. This may allow insurers to cherry-pick those less likely to fall ill for discounts and those more likely for higher premiums. However, by reducing the need to pool risk (as future health will be much more predictable), the use of health insurance may decline rather than rise because those who are at risk are priced out of the market, whereas those who are low risk opt out of general insurance schemes. Insurance is essentially a form of gambling: When the outcomes (the need to claim) become less uncertain, the desire to insure will lessen. Those who (on the basis of genetic testing) expect to need medical coverage become much less likely to be able to buy economical insurance or obtain work (on the assumption they will need extensive future sick leave). Even the normally free-market *Economist* recognizes that this may support the retention of public service medicine to ensure that those who are unable to get coverage are not rendered as an underclass through the availability of better genetic information (*Economist*, 2000). It is therefore no wonder that insurers have been slow to celebrate and extensively utilize such advances.

The development of ever more complex knowledge about the human genome prompts two important questions that exemplify the more general ethical problems with the commodification of information: First, who has control of this sort of information and knowledge (here implying issues of access to and benefit from its medical use)? Second, are there risks that deployment of such information will (re)produce divisions in society between those regarded as genetically low risk and those who, through no fault of their own, find themselves subject to exclusion or *red-lining* (similar to the way that credit scoring has become a mechanism for exclusion from financial services) or are discriminated against on the basis of their likely future health problems? Although in the United States genetic testing of job applicants has been ruled out by federal agencies, in the private sector such practices continue to allow companies to turn down job applicants on the basis of genetic profiling (Lewis, 2000). Meta-technologies, through their ability to codify and collate diverse and complex knowledge resources, have made such enclosure more feasible, and by doing so have enabled a concentration of important knowledge resources. The rise of meta-technologies requires a political response that clearly asserts there are alternatives to the current settlement over these issues.

ENVIRONMENTALISM OF THE NET AND A NEW POLITICS OF KNOWLEDGE

To develop a politics of intellectual property that might start to address problems like those discussed in the previous section, we might look for inspiration to another movement that reacted to the degradation of a global commons: environmentalism. Boyle (1997) notes the environmental movement was deeply influenced by two analytical perspectives: ecology and welfare economics. From ecology it drew the recognition of complex and unpredictable connections between living things in the real world; from welfare economics it drew the recognition that markets frequently (and quite normally) fail to fully internalize the social costs of property use. Crucially, these ideas were not developed in the mainstream of political discourse, but on the margins and then popularized. Currently the nascent global politics of intellectual property is in the first of these stages. This is a problem as in "terms of ideology and rhetorical structure, no less than practical economic effect, intellectual property is the legal form of the information age" (Boyle, 1997, p. 90). There are many crucial areas where IPRs have a profound effect, not just in the realm of the meta-technologies briefly examined earlier (May, 2000), yet they are still treated as arcane, almost nonpolitical legal instruments.

A similar political effort that produced a change in the politics of the environment needs to shift the conception of knowledge to establish a significant role for a broad view of global social utility. The knowledge commons should be established as a global resource owned by humankind collectively, not one that can or needs to be carved up for individual gain. Yet the politics of intellectual property is at a similar stage of (under)development as the environmental movement was in the late 1950s or early 1960s. Despite flurries of interest and outrage at specific issues, two things are notably lacking:

> The first is a theoretical framework, a set of analytical tools with which the issues should be analysed. The second is a perception of common interest among apparently disparate groups, a common interest which cuts across traditional oppositions. (Boyle, 1997, p. 108)

It is here that the frameworks of ecology and welfare economics did the work of constructing a politics of the environment, revealing disturbing conclusions on which a popular movement could be built. Although the patenting of genetic information has been greeted with outrage, as yet there has been little concerted response or linkage to other IPR-related political issues (see Priest and Ten Eyck, chap. 7, this volume). Campaigns like Action Aid's against genetically modified crops have not managed to establish linkages with the more general threat to the knowledge commons represented by meta-technologies.

Therefore, in much the same way that the environmental movement invented the environment, a politics of intellectual property needs to (re)invent the knowledge commons and reembed individuals in the socialized body of knowledge by establishing its social value. By individualizing creation and disembedding it from the social milieu from which all knowledge is drawn, IPRs deny the importance of these commons and reward only a small group of rights holders. Yet there are different suggested solutions to the problem, some of which may represent a reformulation of the existing arrangements, not a radical critique.

For instance, Stiglitz (1999)[6] argues for a compromise that does recognize some of the problems, but limits the actions to the areas of "basic research and many other *fundamental* forms of knowledge":

> Knowledge is a public good requiring public support at the global level. Current arrangements *can* be made to work effectively, but if they are to succeed we must be aware of the dangers and pitfalls. Some countries may try to free ride on others; they may try to capture more of the returns that are available from the use of the global knowledge commons; they may see their self-interest enhanced more by taking out of the global knowledge commons than contributing to it, *in supporting research to design patentable applications rather than supporting basic research.* (pp. 320–321; italics added)

Those who have the audacity to use the knowledge commons to produce patentable innovations without funding basic research (which might be taken to broadly typify the innovation efforts of newly industrializing and developing countries, although they may conduct some basic research) are the threat to these commons, not the companies that own important patents derived from their use or enclosure of the commons.

This suggests the value of the global knowledge commons is already being contested and is amenable to dilution and capture even before the movement has gotten underway. For Falk (2000), this is the danger of "normative cooption" of the idea of a "common heritage of mankind" (such as a global knowledge commons) and is illustrative of the problems when "global civil society is relatively passive and global market forces are mobilized in defense of [sectoral] interests" (p. 327). If nothing else, this reveals the perceived threat such a movement might represent to the knowledge owners. Their concern to defend and consolidate the role of IPRs in the realm of meta-technology represents a triumph for the anti-commons and, as such, is a reversal of the problem Hardin's original argument addressed (Aoki, 1999). In the anti-commons, too many have the right to exclude from use, to withhold (in the commons no one has this right), and this leads to a problem of underutilization of the resource—in this case, knowledge (rather than its overuse and depletion). Underutilization when linked to life- or health-related knowledge, accessed and developed through the use of meta-technologies, must be an issue of global social justice.

SOCIAL JUSTICE?

Claims to justice, suggests Walzer (1983), are a combination of three distributive principles variously weighted by different social groups: free exchange, desert, and need. In another influential analysis, Miller (1976) regarded the elements as rights, deserts, and needs. Whatever the arguments about social justice overall, the way these elements have appeared in discussions of intellectual property is instructive. The notions of free exchange and desert are often appealed to in the justifications of intellectual property through arguments for the need to be able to sell or transfer property to others and the Lockean schema of reward for effort, whereas the idea of rights is a central aspect of both Lockean and Hegelian schemes. However, the only need that appears in these debates is the need for *efficiency* mobilized in the economic or pragmatic justification.

The underlying difficulty is the manner in which needs are identified and, more important, whose needs are given priority where they come into conflict. How the particular need for knowledge efficiency is defined and how this is accommodated within the current intellectual property regime is taken as unproblematic in the prevailing discussion of IPRs. As Walzer (1983) notes, to "argue against dominance and its accompanying inequalities, it is only necessary to attend to the goods at stake and *to the shared understanding of these goods*" (p. 320; italics added). Hence, the politics of the distribution of intellectual property must and will be primarily concerned with defining such property differently. The problem for a new politics of knowledge relating to meta-technologies is that the type of property in knowledge that has become normalized is private property. Therefore, an important aspect of an environmentalism of the internet is to reopen this question and present alternatives to the dominant paradigm of property ownership for knowledge and information. This will not be easy, but neither is it without precedent. As Boyle suggests such a campaign can draw sustenance and inspiration from the environmental movement. For example, George Gerbner's work on the media environment and that of others in the media ecology movement may offer examples of how this analysis could be developed.

Aoki (1999) suggests the traditional argument for the provision of public goods based on their boundlessness, and market failures as regards individual use is another way forward in thinking about the global knowledge commons. Certainly if withholding knowledge and information precludes certain activities, and individuals are unable to enumerate their interests through the market, this seems to make sense. However, currently there are two problems that, although usefully illuminated by this approach, still preclude the immediate establishment of an environmentalism of the internet. First, the history of IPRs has already recognized issues related to the value of a form of knowledge commons and made any intellectual property essentially temporary (albeit for varying periods by type). The problem is not so much the lack of balance between private and public rights over knowledge, but rather that the acceleration (prompted by scientific, technological, and organizational advances) of innovation has disturbed a previously legitimate balance. By leaving the duration of protection unchanged during a period of accelerating innovation, the current legal regime for intellectual property has essentially extended and expanded the advantages for owners. Thus, the field remains open to compromises and partial shifts that do not go far enough yet ameliorate some of the problem: The duration of patent protection might be reduced to recognize the acceleration of innovation and obsolescence while still offering a period of protection to owners. This is a political issue open to argument and compromise; there is nothing natural or final about any particular period of protection.

Second, and perhaps more fundamentally, who would represent the global public interest in protecting the knowledge commons, where might such authority be vested, and how would it become accountable? Could a governing body act as the clearing house for the flow of knowledge to those who needed it within the global system, but lacked the immediate means to pay? This might plausibly be carried out by an agency of the United Nations or an organization affiliated to the WTO. Indeed in a limited way, this is exemplified by the Consultative Group for International Agricultural Research (CGIAR), which, through its scientific centers in some developing states, has collected, analyzed, and recorded biomass resources. By putting these records into the hands of the United Nations Food and Agriculture Organization (FAO) as part of the Plant Genetic Resources for Food and Agriculture program, these bioinformation resources are lodged in the commons (Braman, 2002). A similar body could be entrusted with the wealth of information in the human genome, which while establishing these resources as public allowed considerable potential for the patenting of these resources when combined or synthesized for a clear practical use. Although still potentially problematic, this would at least limit the coverage of patents to novel or innovative (genetic/bio) products provided such criteria were upheld (although current practice would need to be changed). The discussions of the global knowledge commons are therefore one facet of a much larger problem—that of global governance, democracy, and justice.

Intellectual property is not merely some arcane technical legal issue; the emergence of meta-technologies has made information and knowledge an important area of political conflict. It is this rise in importance that make the comparison with environmental politics apposite. As the ownership and availability of knowledge resources become increasingly recognized as a major contemporary problem, the effects of the poor distribution of information and knowledge will become politically sensitive. As emerging meta-technologies affect more and more aspects of our lives, the contemporary shortcomings of intellectual property law will become increasingly apparent. Tim Hubbard from the Welcome Trust commented that, "[The patenting of genes] is like fishing with dynamite. You catch a lot of fish, but you destroy the environment for everyone else" (quoted in Morgan, 2000, p. 4). When these effects start to become more explicit, the justificatory schemata I have explored will become more widely promulgated as the common sense of the political economy of information and knowledge to neutralize new political responses. A political effort will be required to organize against this narrowing of the agenda, and environmentalism offers a good model from which to initially develop a grass-roots politics of intellectual property.

ENDNOTES

1. A number of people have commented on previous versions of this chapter. I especially thank Kurt Burch, Peter Glasner, Roddy Loeppky, Jonathan Nitzen, James Roberts, Harry Rothman, Jem Thomas, and Simon Thompson for their comments. Susan Sell helped me think through many of the issues presented here, and Sandra Braman offered invaluable editorial advice in the final stages of writing.

2. Although this is not presented as the case in the usual account of the history of intellectual property, given that early patents (in Venice and Britain) were predicated on the assumption that the award of a patient carved out particular ownership rights (privileges) from a realm that was implicitly the property of a wider community (the society against which such privileges were awarded), the knowledge commons can be said to have been implicitly recognized when intellectual property was first developed as a legal form. I am working with Susan Sell to further develop this and other aspects of a new international history of intellectual property (see Sell & May, 2001), although my argument here is not dependent on this particular claim.

3. In the United States of America, protection currently is 70 years, which some claim is linked with the Disney Corporation's desire to continue to protect Mickey Mouse through copyright.

4. Article 53(a) of the European Patent Convention excludes from patentability any invention where "the publication or exploitation of which is contrary to morality or ordre public."

5. Copyright has also been used to halt the dissemination of results unfavorable to the funders of research (May, 2000). See also Lievrouw's (chap. 6, this volume) discussion of the problems IPRs present for scholarly communication.

6. During Stiglitz's time at the World Bank, the bank fully supported the activities of a number of multinational corporations whose aim was to secure access to germplasm and other genetic resources found in the developing world. Whether the IPR resulting from such bioprospecting are theft or the appropriate reward for the process of discovery and development of the resources remains a moot point.

6

Biotechnology, Intellectual Property, and the Prospects for Scientific Communication

Leah A. Lievrouw
UCLA

In an article published in *Science* 20 years ago, sociologist of science Dorothy Nelkin (1982) poses this question: "Who should control scientific information?" (p. 704). Considering scientific research broadly, but noting biomedical research as a major focus of interest at the time, she identifies several main tensions among academic scientists, government, and industry.[1] She notes the growing reliance on intellectual property controls (especially patents), instead of more conventional peer review and regulation, to manage and direct the flow of scientific information and interaction. It was becoming more difficult, she argues, to establish standards or uniform principles for intellectual property protection that would balance property interests with the expectations of open and disinterested scientific practice and communication.

By the late 1980s, the issue of intellectual property had become so prominent in discussions of science and technology that sociologist Harriet Zuckerman (1988) reframes the question: "Who has rights of ownership in science, under what circumstances, and how free are they to convey the 'owned' intellectual property to others?" (p. 8). She contrasts what she argues was a traditional Mertonian sense of property rights in science, in which discoveries are shared through open communication and publishing, and recognition and priority are accorded to the first to publish, with property rights in technology, where discoveries are often kept secret to protect the financial interests of investors and ownership is secured by the first to patent. Citing a number of examples, and notably the new university–industry relationships between Hoechst and Harvard, and Whitehead and MIT, she concludes that the growing influence of private ownership in biotechnology was already leading to a modification of a traditional, reputational concept of *ownership* in science.

In retrospect, it is instructive that even two decades ago two prominent sociologists of science—who might be expected to frame scientific knowledge as cultural achievements or even as a public good—should predicate their analyses on an economic concept of scientific knowledge as a commodity or property to be controlled or owned. Even then the biosciences seemed to be leading the way to a new era of blurred and overlapping institutional boundaries and values. Since the 1980s, the trend has accelerated with the widespread implementation of new information and communication technologies (ICTs) for scientific information retrieval, interpersonal interaction, and data management. Today, with one foot in the academy and the other in the market, and supported by a sophisticated infrastructure of networked institutions and information technology, there is no question that biotechnology poses some of the most important challenges to the conventions and norms of scientific practice, and therefore to the development of scientific knowledge.

If, as Nelkin, Zuckerman, and others claim, the gaps between science and the market were narrowing in the 1980s, it can be argued that in biotechnology today they have effectively disappeared. Along with them has gone whatever tension existed between property rights and competitive secrecy, on the one hand, and scientific ideals of open communication and disinterestedness, on the other. Intellectual property and the conversion of research findings to lucrative products have become governing considerations for both private firms and universities (part of what legal scholar Jessica Litman [1994] calls the "intellectual property epidemic"). Questions about the control of scientific information in biotechnology are still widely debated, but to an increasing extent are becoming moot in the wake of a relentless drive for capital investment, patents and licensing, and marketable products.

The control of information in biotechnology today is a matter of property ownership and financial stake (as May details in chap. 5, this volume). It clearly lies with those who hold and commercially exploit intellectual property rights and markets in information, whether they are private firms that conduct and fund research or the major research universities and institutes that collaborate with industry or hold patents derived from publicly funded research. As often as not, secrecy and competitiveness have become the norm in both private sector and academic biotechnology research. Scientists still share their knowledge, but a host of recent cases suggests that free and open exchange tends to be limited (especially by contractual mechanisms like confidentiality agreements or material transfer agreements) to more tightly bounded problem groups or research teams rather than conducted through traditional disciplinary or institutional channels.

Because private sector and academic science have merged so extensively in biotechnology, the field provides an ideal context for a retrospective exploration of the development of scientific knowledge, particularly the consequences of privatization and expanding intellectual property claims for the communication relationships and practices that produce and share that knowledge. The present discussion has three main parts. The first is a brief historical overview of the development of knowledge in biotechnology from its craft-based origins to today's computational simulations and modeling. Particular attention is paid to the changing institutional contexts of biotechnology that have helped to shape, and have been shaped by, the communication practices and patterns among researchers in the field.

In the second part of the discussion, several themes or issues in the historical overview are elaborated that have been particularly consequential for scientific communication in the field. These include changes in the nature of knowledge and its products over time; shifts in the central problems addressed by the field and its modes of knowledge production and use; and the changing relations among various groups, institutions, and interests.

In the third and final part, I summarize several key features of scientific communication in biotechnology research today that have developed in this changing context. To varying degrees, they depart from the norms and practices that scientists have traditionally espoused. They include what I call the retreat from publication; publication bias; the erosion of peer review; and growing constraints on informal, interpersonal interaction among researchers. I conclude with some brief comments about the implications these changes may have for scientific and scholarly communication more generally.

THE GROWTH OF KNOWLEDGE IN BIOTECHNOLOGY: A BRIEF HISTORY AND BACKGROUND[2]

Although definitions are perpetually debated, one widely quoted description of biotechnology is "the application of scientific and engineering principles to the processing of materials by biological agents to provide goods and services" (Bull et al., 1982, p. 21). However, the principles involved have developed in a variety of ways, ranging from the tacit knowledge and cultural heritage of craft workers such as brewers and cheese makers to the speculative induction of today's computational modelers. Broadly speaking, historians and other observers of the field have described three main eras of biotechnology research: classical, modern, and new biotechnology. Each can be characterized in terms of the central issues or problems that motivated research, its distinctive modes of knowledge production and use, and its institutional forms and relationships. (Major points from the following sections are summarized in Table 6.1.) Obviously such a scheme can oversimplify the scope and diversity within a field, but it can also help frame important events and developments into a basic and brief introduction.

Classical Biotechnology

As a body of knowledge, it is often noted that biotechnology has ancient roots in cultural practices developed across many regions and peoples. Humans have known for millennia how to make wine and beer, sake and shoyu, and cheese and vinegar; how to make leavened bread; and how to cure meats and tan leather. Although it was not known until the 19th century that microorganisms were involved, techniques of fermentation, brewing, leavening, tanning, and so on have long been understood, systematically practiced, and have become part of cultural heritage throughout the world.

This period, from ancient times through about the middle of the 19th century, has been called "classical biotechnology" (Organization for Economic Cooperation and Development, 1989, p. 4). Knowledge existed mainly in the form of practical skills and experience, which were often tacitly understood and passed along in apprentice relationships. Later knowledge was shared informally among members of skilled trades or guilds, such as those for brewing, baking, or cheese making. Eventually, as trades grew into industries, these groups organized and sponsored their own workshops and laboratories where they could experiment with new techniques or refine old ones, thereby expanding and formalizing what was essentially a body of craft knowledge.

In 1697, "at the very end of the century of the 'scientific revolution,'" Georg Ernst Stahl coined the term *zymotechnia*, or zymotechnology, to denote the study and applications of fermentation (cited in Bud, 1993, p. 8). Now considered one of the founders of scientific chemistry, Stahl hoped to establish fermentation as one of the pillars of the new field. His book, *Zymotechnia Fundamentalis*, was also "a bid for intellectual authority over related commercial processes" (Bud, 1993, p. 9), and therefore an attempt to further codify and formalize the craft knowledge of fermentation within a more scientific and academic framework.

However, over the next century, fermentation chemistry was displaced by organic chemistry as the intellectual and research focus of the field, not least because of the role that organic chemistry played in the rapid growth and dominance of the German chemical industry. By the early 19th century,

TABLE 6.1
The Development of Knowledge in Three Periods of Biotechnology Research

Variable	Classical Biotechnology (Ancient society to mid-19th c.)	Modern Biotechnology (Mid-19th to mid-20th c.)	New Biotechnology (Mid-20th c. to present)
Nature of knowledge	Skills, experience Cultural heritage Tacit knowledge	Hypothetico-deductive Scientific/disciplinary Generalizable, theory-based	Inductive, post hoc theory Simulations, models Pattern-finding
Modes of knowledge production and use	Craft, description Trial-and-error application	Empirical observation Controlled experimentation Publication, public presentation	Computation, modeling Synthesis, simulation Secrecy, competition
Central problems	Small-scale craft production Practical application Imitating nature	Industrial production Improved quality of life Harnessing nature	Control and manipulation of natural processes Remaking nature
Knowledge products	Goods, commodities Practices	Mass-produced products Publications Reference collections of organisms	Patents, licenses Combinatorial libraries Sequence databases
Institutions/players	Guilds, craft workers Trades Early scientific societies	Specialized private firms Private research institutes Agricultural, land-grant colleges Scientific/engineering disciplines	Research universities, labs Public funders and regulators Markets, investors Faculty-entrepreneurs Diversified private firms

"[c]ompared to the detailed development of organic chemistry, fermentation chemistry was a marginal, empirical, and messy province" (Bud, 1993, p. 14).

Modern Biotechnology

However, in 1857, Louis Pasteur showed that lactic acid-producing bacteria were responsible for souring beer and wine, thereby ushering in the modern era of biotechnology that continued through the middle of the 20th century (Organization for Economic Cooperation and Development, 1989, p. 4). He and his colleagues put "fermentation at the centre of [the field's] concerns . . . [Pasteur] constructed a new scientific discipline based on his understanding of microbes, 'microb-iology' " (Bud, 1993, p. 14).

Once the role of bacteria, yeasts, and other organisms in chemical processes was recognized, zymotechnology enjoyed a revival, particularly in private industry. Firms were eager to exploit the growing understanding of the chemical products of microbial metabolism and to produce these products (e.g., acetic acid, lactic acid, alcohol, methanol, glycerol, acetone, and butanol) in industrial quantities. Industrial chemists began to interact with those involved in the younger biological disciplines of microbiology, bacteriology, mycology, and botany.

At the same time, several prestigious, industry-affiliated private laboratories and research institutes were founded during the late 19th century, including the German Institüt fur Gärungsgewerbe, the Carlsberg Institute and Alfred Jorgensen's Fermentology Laboratory in Denmark, and the Institute of Fermentology (later the Zymotechnic Institute) in Chicago. In Denmark, the journal *Zymotechnisk Tidende* began publication in 1885, and in the 1880s, the *American Chemical Review* was retitled *The Zymotechnic Magazine.*

As microbiological and biochemical processes were elucidated, industries increasingly viewed the knowledge gained from basic research as a necessary source of innovations. They formed closer relationships with, and established industrial labs with, European institutions of higher education such as the British School of Malting and Brewing, founded at Birmingham University in 1899, and its Scottish counterpart at the Andersonian Institution (later Strathclyde University). In the United States, industries allied themselves with land grant universities.

By the early 20th century, microbiologists had begun to recognize the significance and applicability of their work beyond a few key industries. Zymotechnology, which had been more a creature of industry than the academy, began to shift away from a primary focus on fermentation and toward microbiology more broadly, the use of scientific method, and the discovery and application of scientific principles.

Reference collections of microbial cultures were established at the Technical University of Prague as well as in Berlin and Holland. The number of academic research publications grew steadily, including journals like the *Zeitschrift für Technische Biologie*. The term *biotechnology*, which captured this new sensibility, was introduced as early as 1917 by the Hungarian researcher Karl Erecky, who defined it as "all such work by which products are produced from raw materials with the aid of living organisms" (Erecky, 1919; cited in Bud, 1993, p. 27).

Significantly, throughout the modern era of biotechnology the field was regarded by intellectuals and the public alike as the benign and progressive face of technology. In the second and third decade of the 20th century, it contrasted starkly with the dirty, dangerous, and dehumanizing mechanical and industrial methods and technologies that had been demonstrated with such terrifying effectiveness during World War I. For example, industrial microbiologists who learned how to ferment acetone and butanol for use in explosives and synthetic rubber during World War I also developed techniques for the mass production of citric acid and other flavorings used in food products, including soft drinks (Primrose, 1991). Waste water treatment and municipal composting methods were introduced and remain among the most economically and socially significant applications of biotechnology worldwide (Smith, 1985).

The sense of biotechnology as *chemurgy*, or chemistry at work, was slowly overtaken by the idea that biology and engineering might be harnessed together to solve human problems and develop techniques to improve the quality of life for more people. Early proponents of biotechnology idealized it as organic, spiritual, and ethical. This *bioaesthetic* perspective was closely identified in the 1930s with scientists and social critics such as Lewis Mumford, Julian Huxley, Lancelot Hogben, and J. B. S. Haldane.

Biotechnology also promised to liberate humans from the random workings of nature. "The Darwinian image of the survival of the fittest was portrayed as wasteful and derogatory of the human being" (Bud, 1993, p. 57). Improvements could be achieved by changing either the organism or its environment; this basic principle of eugenics was applied in the new and thriving fields of hygiene, food science and nutrition, epidemiology and "social medicine," sociology and human economy, birth control, and feminism. (Of course eugenic principles were also appropriated and misused later as part of the National Socialist political program in Germany.)

Perhaps the most consequential product of late-modern biotechnology, however, was the discovery of the antibiotic properties of substances produced by microorganisms such as *Penicillium* and *Streptomyces* species. Antibiotics brought the medical applications of biotechnology to center stage and initiated an era of unprecedented research growth in the pharmaceutical industry. Methods were developed for culturing desirable microbes in mass quantities under aseptic conditions, thus preventing contamination by unwanted organisms (Primrose, 1991; Smith, 1985). Microbiologists learned how to isolate, select, and breed just those strains of organisms that seemed most useful or productive, and they began to screen thousands of naturally occurring organisms to find those that produced pharmacologically active agents, such as antibiotics, enzymes, or amino acids. They developed techniques for large-scale cultivation of animal cell cultures as part of the mass production of vaccines.

In the context of these successes, U.S. universities established the first academic programs in biological engineering (at MIT in 1939) and biotechnology (at UCLA in 1947). UCLA received a major foundation grant in the 1950s to develop an undergraduate bioengineering curriculum. Other prestigious institutions such as Carnegie-Mellon University also began to offer courses. In Europe, the Royal Swedish Academy of Engineering Sciences created a new *bioteknik* section in 1942, and the Biological Engineering Society was organized in Britain in 1960.

As represented in these programs and organizations, the domain of biotechnology in the 1940s and 1950s had clearly expanded beyond its traditional roots in biochemistry and microbiology. The new programs incorporated studies of man-machine systems, bionics, human factors engineering, ergonomics, and medical electronics as well as more traditional areas like physiology, microbiology, and biochemistry. These mid-century programs also bore little resemblance to biotechnology as the term is popularly understood today. Yet they constitute an interesting turning point or knowledge link in the transition from *modern* biotechnology, which focused on living organisms and their products, to *new* biotechnology, which reframed those organisms and products in terms of information.

New Biotechnology

Up to the mid-20th century, modern biotechnology was grounded in the increasingly sophisticated craft knowledge of industrial production, the maturing scientific knowledge of fermentation chemistry and microbiology, along with the emerging specialty of bioengineering. Its main purpose was the "biological production of chemicals" (Bud, 1993, p. 45).

However, beginning in the 1950s, the field underwent a radical shift in theory and emphasis in the wake of pioneering developments in genetics and molecular biology, notably Watson and Crick's description of the structure and function of DNA (Bud, 1993; Kenney, 1986; Watson, 1968/1980). This latest incarnation of the field is called the "new biotechnology" (DeForest, 1988, p. 5; Organization for Economic Cooperation and Development, 1989, p. 4; Office of Technology Assessment, 1988).

From Microbiology to Genetics. In the 1960s, researchers' attention turned from understanding the metabolism and biochemical pathways of intact, naturally occurring organisms to the manipulation of genetic material as a means to produce new organisms with desirable traits (such as the ability to produce a particular antibody or other protein). The long-established skills and processes involved in selecting and culturing naturally occurring microorganisms for their useful traits were still taught, but "relegated to obscurity" (Bud, 1993, p. 164).

In the early 1970s, two developments accelerated the shift toward molecular biology and genetics. In 1973, Stanley Cohen of Stanford University and Herbert Boyer of the University of California–San Francisco spliced genetic material from a frog into a strand of bacterial DNA, transferring the frog's traits into the resulting bacterial cell lines. Their technique for making recombinant DNA immediately transformed the biotechnology research agenda. It has been called "the single pivotal event in the transformation of the 'basic' science of molecular biology into an industry" (Kenney, 1986, p. 23). Today the products of recombinant DNA procedures include growth hormones; insulin; tissue plasminogen activator (TPA), which is used to dissolve blood clots; and the anticancer agent interferon (Ducor, 1998; Primrose, 1991; Smith, 1985).

In addition to recombinant DNA, cell-fusion techniques (which involved breaking down the walls of different cells so that their genetic material might intermingle) were developed to create new varieties of plant and animal cell cultures. The resulting cell culture would be cloned and reproduced indefinitely from the single new type of cell to produce particular proteins in quantity. For example, hybridomas fuse spleen cells (taken from an animal that has been exposed to a particular antigen, and so produces antibodies for that antigen) with tumor cells. The resulting hybridoma cells can both produce large amounts of easily purified antibodies and have the almost limitless growth potential (or "immortality") of a tumor. Today hybridomas and other cell cultures are the primary source of important agents like monoclonal antibodies.

The "wedding with genetics" (Bud, 1993) was not completely blissful, however. Early on strains emerged between applied microbiologists and industry scientists, on the one hand, and the molecular biologists entering the field, on the other. In February 1975, the watershed Asilomar Conference brought together about 150 participants to discuss the risks of, and to establish rules for, conducting recombinant DNA research. Asilomar embodied biotechnology's new intellectual agenda, recently described by anthropologist of science Paul Rabinow (1996) as its

... potential to get away from nature, to construct artificial conditions in which specific variables can be known in such a way that they can be manipulated. This knowledge then forms the basis for remaking nature according to our norms. (p. 20)

However, the meeting also embodied an "enduring tension between microbiologists and molecular biologists," which would continue to arise in academic curricula and research agendas in the years that followed. All of the conference organizers and most of the participants were top academic molecular geneticists (Kenney, 1986; Krimsky, 1982). They had few affiliations with the microbiologists who attended, many of whom worked in industry. The outnumbered microbiologists accused the molecular biologists of being too focused on subcellular processes and "lacking a feel for the organisms" (Bud, 1993, p. 176). On a BBC TV broadcast after the conference, a microbiologist who had attended derided "the average molecular biologist, whose manipulation of bacteria chills the blood of anyone accustomed to handling pathogens" (cited in Bud, 1993, p. 176). Nonetheless, the conference would be hailed later as the "beginning of the biotechnology age" (Bud, 1993, p. 175).

Regardless of the initial disciplinary tensions, by the late 1970s, genetics completely dominated the biotechnology research agenda and had become the field's main attraction (and concern; see Murdock, chap. 9; and Priest & Ten Eyck, chap. 7, this volume) for investors, established industries, regulators, and the public, as well as in the academy. According to one recent study, 64% of news stories about biotechnology carried in major U.S. print media between 1975 and 1979 dealt with the theme of DNA research, whereas only 16% dealt with microorganisms (Nisbet & Lewenstein, 2001). The lay public and biotechnology researchers alike grew to see the field almost exclusively in terms of genetics and, in particular, DNA. Indeed aside from occasional obligatory nods to ancient brewing or leather-tanning practices, most contemporary surveys of the field have tended to ignore the premolecular biology history of the field entirely (see e.g., DeForest et al., 1988; Kenney, 1986; Rudolph & McIntire, 1996; Zweiger, 2001).

The Biotechnology Enterprise. Despite the public's fascination, biotechnology research remained a relatively small and arcane specialty through most of the 1970s. The majority of research was federally supported and conducted in academic labs—a pattern that would continue well into the 1980s. In 1975, the National Institutes of Health (NIH) funded just two biotechnology projects at a cost of $20,000 (although NIH support increased dramatically in 1976 to $15 million, distributed across 123 projects; Rabinow, 1996). Until about 1979, only four private biotechnology firms were in operation (Biogen, Cetus, Genentech, and Genex; Bud, 1993); but around 1980, several key economic and policy factors combined that helped to recast the field as a market-driven enterprise.

The first factor was legal. In the 1980s, in an effort to speed up commercialization, the U.S. Congress passed the Bayh–Dole Act, which gave universities the right to patent discoveries arising from federally funded research. Previously, universities and faculty members had only been able to patent inventions resulting from university-supported or privately funded research, and most major research institutions filed few patent applications.

But the Act essentially pushed universities into the market, requiring them to report any potentially patentable findings stemming from federally supported research projects; otherwise the patent rights would revert to the government (Rabinow, 1996). Universities could now own and license the rights to the inventions and discoveries made by taxpayer-supported faculty and student research. Institutions immediately recognized the revenue potential of patenting. Directly after the Act went into effect, for example, UC San Francisco and Stanford jointly applied for and received a patent for the Cohen–Boyer method for making recombinant DNA. By the time it expired in 1997, the patent was estimated to have earned more than $150 million (Benowitz, 1996).

Estimates of increased university patenting activity vary widely, but they all indicate the significance of the change in law. For example, the Office of Technology Assessment (OTA) reported that university patent applications rose by 300% between 1980 and 1984 (Rabinow, 1996). *Scientific American* reported that fewer than 500 patents were awarded to universities in 1980; by 1994, that figure had risen to well over 1,500 patents per year (Gibbs, 1996). *Nature* stated that 40% of patents on human DNA sequences in 1995 were awarded to universities—about double the proportion of a decade earlier (Thomas, 1997). *Le Monde Diplomatique* claimed that the number of patents produced by universities increased 20-fold between 1981 and 2001 (Warde, 2001).

Not only were universities being encouraged to increase their patenting activities; patents were also being awarded for new kinds of "inventions." In 1980, in the watershed *Diamond v. Chakrabarty* case, the U.S. Supreme Court ruled that new genetically modified organisms (in this case, a bacterium modified to digest crude oil) could be patented. The court set aside previous interpretations of patent law that had allowed the products of biological organisms (such as antibiotics or enzymes) to be patented, but not the organisms (except plants) because they were "products of nature." In 1987, the Commissioner of Patents and Trademarks issued a statement that the Patent and Trademark Office would henceforth consider all "nonnaturally-occurring," multicellular organisms, including animals, to be patentable, although humans were excepted (Ducor, 1998). In 1988, the Supreme Court, following the doctrine established in *Diamond v. Chakrabarty* (1980), awarded U.S. Patent No. 4,736,866 to Harvard University for a transgenic mouse, ending a string of legal challenges to the patentability of ever more sophisticated animals.

The second factor that helped move biotechnology in this new direction was financial. Once patent rights in biotechnology products were expanded, investors discovered biotechnology as a new high-tech field that appeared to have the same payoff potential as information technology. Previously, biotechnology had been considered more as a body of techniques whose main value was their application in other fields, such as pharmaceuticals, agriculture, or industrial chemistry, than as an industry in itself. Indeed firms from all those industries had conducted biotechnology research or had supported academic research.

But venture capitalists and biotechnology researchers who had seen the spectacular performance of the Silicon Valley and Route 128 startups envisioned "a new biology-based industry, on the model of information technology," with academic scientists serving as entrepreneurs. Biotechnology, information technology, and materials science were often cited together in the 1980s as a trinity of high-tech "metaindustries" with a wide range of potential uses and broad-based economic and social prospects (Bud, 1993; Organization for Economic Cooperation and Development, 1989). The brokerage E. F. Hutton, led by pharmaceuticals analyst Nelson Schneider, was so enthusiastic about the parallel between biotechnology and information technology that it began publishing the first investment newsletter devoted entirely to the area. In 1983, Hutton also organized a research and development limited partnership, California Biotechnology, to produce and market the biotechnology products of startup firms with links to universities (Bud, 1993, pp. 183–184; Kenney, 1996). Even "regulators . . . seemed to have been successful in finding a consensus that biotechnology, properly controlled by responsible scientists, could be seen as a latter-day information technology" (Bud, 1993, p. 214).

In what Kenney (1986) calls the "gold rush atmosphere" of 1979 to 1981, a few universities launched startups of their own—notably Michigan State University. Most avoided the appearance of conflict of interest by continuing to manage and license their patents through separate but university-controlled brokers such as University Patents or the Research Corporation, as they had previously. Some companies provided financial support for whole laboratories or departments in exchange for ownership interest in the research produced by faculty and students (e.g., Hoechst and the Massachusetts General Hospital; Monsanto and Washington University; and the Whitehead Institute and MIT [Office of Technology Assessment, 1984]).[3]

More often, faculty entered into individual relationships with private firms. Their roles ranged from service as paid members of scientific advisory boards or consultants, to equity positions in companies (in some cases large enough to be reported in a firm's prospectus), to employment in management positions, to seats on a firm's board of directors. Critics of this development saw

[T]he appearance of a new academic type: the professor-entrepreneur who uses his academic affiliation as a launching pad for lucrative ventures ... [with a] tendency to privatise revenues and socialise expenses (through the use of university administrative resources as well as "free" student labour). (Warde, 2001, p. 1)

Krimsky, Ennis, and Weissman (1991) define university faculty with these types of relationships as *dual-affiliated biotechnology scientists* (DABS). They surveyed 832 American academic life scientists (plant pathologists, microbiologists, geneticists, and biochemists) between 1985 and 1988 who reported having 927 links with private biotechnology firms. Universities with top biotechnology research programs had the highest proportion of DABS; about 31% of MIT life scientists had industry links, compared with about 19% at Stanford and Harvard and an average of 14% of the relevant faculty at five major University of California campuses. Krimsky et al. (1991) point out the possible negative consequences of such pervasive relationships for traditional activities like peer reviewing, noting that some journals were having difficulty finding disinterested reviewers for certain areas of research.

The third factor in the push to transfer university research to private ownership was a host of new tax laws. In 1980, changes in the tax code created research and development limited partnerships (RDLPs) to provide "special tax shelters and high investment income" for investors in university-industry research projects. The Economic Recovery Tax Act of 1981 gave a "25% tax credit to 65% of a firm's payment to universities to support basic research" (Krimsky, 1988, p. 36).

However, the initial burst of venture capital and the proliferation of IT-style startups ended quickly. The number of biotechnology firms spiked in 1981. Only four biotechnology firms were in business through most of the 1970s; 26 new companies started in 1980 alone, and over 40 more started in 1981. Yet in 1982, the number dropped back to 22. By 1987, after rounds of mergers and acquisitions, the number of truly independent biotechnology companies had returned to near mid-1970s levels, although many small startups survived when they were bought and subsequently operated as separate firms by large corporations (Bud, 1993; Kenney, 1986; Krimsky et al., 1991).

In part, the boom and bust was due to the realization among investors that biotechnology research could not possibly generate the kinds of short-term returns that many had grown to expect from high technology. It also became apparent that, despite the drive by investors, the OECD, and others in the 1980s to cast biotechnology as a metaindustry like information technology, pharmaceuticals and agricultural applications would remain the most lucrative areas for the foreseeable future. Drug and agribusiness firms, which were undergoing multinational mergers and consolidations with chemical companies and each other, became the primary source of capital for biotechnology ventures. The smaller firms and their intellectual property became acquisition targets when the honeymoon in the markets ended (Bud, 1993; Kenney, 1986). To date biotechnology has not fulfilled the extravagant expectations of the 1980s in terms of either sales or return on investment. Rather than emerging as an independent metaindustry, biotechnology has remained a supporting player, principally supplying new tools for research and development in agribusiness and pharmaceuticals.

Biotechnology and Information Technology. However, perhaps no development influenced biotechnology as profoundly between the 1970s and the 1990s as the accelerating rate of innovation and growth in information technology, and, more specifically, its applicability to biotechnology research. "It is probably no accident that the rise of the science of genomics has almost exactly paralleled the rise of the science of computing" (*Economist*, 2000, p. 4). As in other research specialties, information and communication technologies were increasingly employed in biotechnology to deal with growing problems of information retrieval, communication, and data management—the three activities that comprise *bioinformatics* in its broadest and original sense as developed in information science (Saracevic & Kesselman, 1993).

Information retrieval includes the selection, collection, organization, management, access to, and retrieval of all types of documents and records. The postwar "big science" boom in American higher education and publicly funded research (Price, 1963)—of which biotechnology and bioengineering were an integral part—produced a corresponding information explosion in the sciences and academic publishing. The new wave of research literature created a crisis in information organization, management, and retrieval, and computer-based automated document indexing and retrieval systems were built to help manage the volume of materials. By the 1970s, most major research libraries ran their own online public access catalogues and subscribed to an assortment of public and private databases of research literature (e.g., DIALOG, MEDLINE, LEXIS). Such systems and services brought diverse and arcane documents together in easily searchable resources that vastly expanded the scope and availability of research available to most scientists.

Meanwhile as the sciences (including the biosciences) proliferated into dozens of subspecialties and research fronts with members scattered all over the world, scientists employed new telecommunications and computing technologies to maintain and extend their networks of contacts with colleagues, students, administrators, regulators, funding agencies, industry sponsors, and so on. "Scientific communication" involved far more than the collection of literature, exchange of preprints, or presentation of conference papers. By the 1980s, written correspondence among researchers and labs was routinely augmented by telephone conversations, conference calls, faxes, and e-mail (Lievrouw, 1986). The fledgling ARPANET had evolved into the internet, linking major American research universities with other institutions internationally (Lievrouw, 2002).

However, data management (including collection, organization, transfer, analysis, and reporting) seemed to pose the biggest challenge of all. Molecular biologists and geneticists migrating into biotechnology research in the 1960s and 1970s brought their tools with them, including gel electrophoresis, DNA sequencing, spectroscopy, cloning, and gene mapping. They developed algorithms for aligning, comparing, and modeling DNA sequences. These techniques quickly created an "explosion of sequence data" in need of analysis and management (*History of Bioinformatics*, 2001).

Therefore, in the 1980s, as biotechnology evolved from an academic research front into a high-tech enterprise, it also became inextricably tied to information technology. New firms and research groups formed to deal with the data organization and analysis problem generated by sequencing, which had "become routine for molecular biology laboratories" (Zweiger, 2001, p. 34). The introduction and spread of sequencing machines were called the equivalent of the Industrial Revolution in biotechnology (*Ibid.*, p. 59), transforming labs into "sequencing factories" (*Ibid.*, p. 80).

Major sequence databases were established, including Swiss-Prot and the European Molecular Biology Laboratory (EMBL) databases in Europe and, notably, GenBank, operated by the National Center for Biotechnology Information, a division of the National Library of Medicine (Richon, 2001; Zweiger, 2001; see also May, chap. 5, this volume). GenBank rapidly became an important international repository of sequence data. Researchers shared their data eagerly; some peer-reviewed journals even required authors to submit their annotated data to GenBank as a condition for publication (Zweiger, 2001). However, public databases tended to be redundant and "noisy." They were soon challenged by private, for-profit firms that collected and sold specialized data libraries, designed software to analyze or visualize sequence data, or performed statistical analyses and sold the results.

Amid this "sea of biological information" (Zweiger, 2001, p. 189), the push began for a project to completely characterize the human genome. In 1985 and 1986, meetings were held to set the agenda and enlist the interest of the U.S. Department of Energy (DOE), especially to redirect DOE funding from other politically unpopular projects like the superconducting supercollider to the human genome project. In 1987, the DOE set up three research centers, and the National Institutes of Health (NIH) secured additional appropriations for human genome research. The Human Genome Project (HGP) was officially launched in 1990. The private firm Celera (owned by Applied

Biosystems, the world's largest maker of sequencing machines) quickly entered the field to compete with the HGP and created imbalances in information sharing between public and private projects (*The New York Times*, 2000). The initial drafts of the genome were announced simultaneously by the HGP and Celera in 2000, years ahead of the original project schedule.

Biotechnology had become a rich source of problems for applied computer science and statistical research—an area of study dubbed *computational molecular biology*. Sequencing was cast as an important class of problems, and software developers sought to "develop computational methods for inferring biological function from sequence alone" (*History of Bioinformatics*, 2001). The term *genomics* was adopted to refer to "the scientific discipline of mapping, sequencing and analyzing genes" and as the title of a new journal (Richon, 2001).

At the same time, the meaning of *bioinformatics*, at least within biotechnology, narrowed radically to "organizing, classifying, and parsing the immense richness of sequence data" (*History of Bioinformatics*, 2001; Richon, 2001). Characteristically, most current accounts of bioinformatics make essentially no reference to the broader body of information science research on information organization and retrieval, and on scientific communication, which was conducted prior to the sequencing data glut (see e.g., Gibas & Jambeck, 2001). Zweiger (2001) at least alludes to it by noting that the National Center for Biotechnology Information (NCBI) was established in 1982 as part of the National Library of Medicine (NLM), which "already had over a decade of experience with MEDLINE, a database of articles from medical research journals" (p. 40). Still Zweiger's aside vastly understates the pioneering role of the NLM in the design and building of online databases and information retrieval and its early expertise in administering globally distributed scientific information resources.

Genetics, Information, Property. As new biotechnology has become ever more dependent on and interwoven with information technology and has promulgated the new, restrictive sense of bioinformatics, one important outcome is that the metaphorical relation between genetic matter and information seems to have collapsed (see Braman, chap. 4, Ritchie, chap. 2, this volume). The physical, material substances of living organisms and information are now routinely treated as an isomorphism as when the *Economist* (2000), in a recent feature on human genome research, remarked that genomics has "literally" become an information science (p. 4).

Certainly, this informational shift within biotechnology was not an isolated phenomenon. The field was part of a broader cultural shift from the 1960s forward toward viewing more aspects of the material world (including genetic material) as information and seeing information as a type of commodity to be produced, owned, traded, and consumed like other goods.

New electronic media and information systems proliferated as publishing, broadcasting, and telecommunications merged with computing. Throughout society, *information* became a popular metaphor for human action, cultural production, social relations, and indeed for the material world. This idea eventually crystallized in accounts of the *information age* or *information society* (Bell, 1973; Schement & Lievrouw, 1987; Webster, 1995).

Systems theory and information theory were important influences in the biosciences generally. Yet in genetics, where genetic material was described from the start as *code* or *language*, and the processes of reproduction and inheritance as "information transfer" or "translation," the conflation of genetics and information, and of information and property, was profound and fundamental. From visible organisms and tissues, to subcellular and molecular structures and processes, biological phenomena have been reconstrued as information. For example, in the preface of a recent book on patenting in biotechnology, the author stated that "modern biotechnology aims to exploit the genetic information *present* in living organisms" and genes are merely the "*embodiment*" of genetic information (Ducor, 1998, p. v; italics added). This tendency to treat biological matter and processes as abstract, manipulable symbol systems may have seemed natural to molecular biologists, but it also may have helped sever the remaining ties with the older generation of microbiologists with their old-fashioned "feel for the organisms."

In his recent book, *Transducing the Genome*, Zweiger (2001) attempts to capture the contemporary informational Zeitgeist in biotechnology: "Biology is being reborn as an information science, a progeny of the Information Age . . . Molecules convey information, and it is their messages that are of paramount importance" (p. xi). Therefore, Zweiger equates genetic material with information and says that biology has been radically and permanently transformed by the recording and compilation ("transduction") of genetic data into electronic databases.

For example, instead of screening organisms or compounds directly, "combinatorial molecular libraries" (Ducor, 1998, p. 105) containing exhaustive lists of possible molecular combinations are now produced computationally and evaluated at unprecedented speed. Mutations that would otherwise take hundreds of generations to appear naturally or *in vitro* are simulated. These new, manufactured genetic combinations are screened for possible receptors (binding properties with particular compounds) or to see whether they might produce other useful proteins; if so, patents are immediately applied for. Such a computational approach is also being employed to create model cell lines or even to simulate whole-organism models for testing new drugs (Zweiger, 2001, pp. 215–216).

Zweiger argues that with the help of information technology, the old paradigm of biology, based in the hypothetico-deductive model of science and the reductionistic view that genes determine or predict traits in the organism, has been overturned in favor of a new inductive approach. Zweiger's view was presaged in 1993 in the introduction to a special issue of the *Annals of the New York Academy of Sciences* on trends and policy in biotechnology: "[T]he data explosion [is] currently shifting the paradigm of biological research from experimentation to concept formulation" (Falaschi & Tzotzos, 1993, p. x). Data mining, modeling, knowledge discovery in databases (KDD) techniques, multivariate statistics (e.g., discriminant, factor, and cluster analysis; multidimensional scaling), and self-organizing systems theory have moved biotechnology toward an inductive, pattern-finding style of research and post-hoc theorizing.[4]

THEMES IN THE DEVELOPMENT OF BIOTECHNOLOGICAL KNOWLEDGE

Obviously the full and complex story of biotechnology research cannot be summarized in a few pages. The necessarily brief synopsis offered here has concentrated on major developments that have affected the growth of knowledge in the field over time. Yet from even this brief treatment, several points or themes emerge that have had important consequences for scientific communication.

The Nature of Knowledge and Its Products

First, it is clear that the nature of biotechnological knowledge has been transformed over time. The craft knowledge and skills of winemakers, tanners, and bakers was gradually systematized, first as *zymotechnia* and early fermentation chemistry, and then into the scientific disciplines of microbiology, bacteriology, mycology, and biochemistry. Later, molecular biology and genetics/genomics came to center stage. Experiential knowledge was subsumed by hypothetico-deductive forms of knowledge that gave scientists the foundation for prediction and generalization. More recently, due to the development of new computational tools, these earlier forms of knowledge have arguably been superseded by inductive knowledge derived from exploration and "data mining," or even speculative fabrications and simulations of genetic data and structures that have come to be referred to as *in silico* biology (Economist, 2001b, p. 30; Zweiger, 2001, p. 83).

By the same token, the products of these various types of knowledge have changed. Whereas early craft workers and artisans sought to produce more appetizing cheese or sake, modern industrial scientists used systematic knowledge to develop large-scale mass production processes for dyes, acids, or drugs. In the modern era, academic scientists established collections of microorganisms as study resources for colleagues. They created new outlets for publishing their findings, hoping to spur further research and expand knowledge in their fields.

New biotechnology has also yielded a number of practical applications, including genetically modified (GM) foods, mass-produced therapeutic hormones and other proteins (e.g., insulin and TPA), and analytical tests and techniques (e.g., DNA testing). However, to date the tangible products of biotechnology research (and the revenue from them) have been far outweighed by essentially intangible and informational products: patents, licenses, and other specialized rights agreements, and new firms and interinstitutional relationships. Much as modern scientists created reference collections, today's biomedical researchers have established vast databases and libraries of naturally occurring and synthesized gene sequences. Unlike the reference collections, however, many of the new databases, as well as the information they contain, are strictly proprietary.

Central Problems: Modes of Knowledge Production and Use

In many ways, the main challenge for premodern craft workers and tradesmen was producing desirable products in enough quantity and of sufficient quality and consistency to supply their local customers. We might say that their central problem was the imitation and use of natural processes to do such things as raise bread or cure meats.

Modern biotechnology, however, was about scaling up and refining these processes for the demands of industrial production and mass consumption, as for example, in the introduction of the stirred-tank fermenter for the mass production of antibiotics (Primrose, 1991). As with other applications of industrial mass production, a major objective of modern biotechnology was improved quality of life in the form of widely available and affordable consumer goods, improved hygiene and health, and so on. We might say that the central problem of the modern era was harnessing and controlling natural processes on a large scale.

In the new biotechnology era, in contrast, nature is not merely imitated or harnessed; it is in fact *remade*, as Rabinow (1996) pointed out, "according to our norms" (p. 20). Research is notable for its use of ICTs to capture and analyze data, to automate or simulate real biological processes, and, crucially, to design and model entirely new processes and even life forms. In a certain sense, information technologies have had much the same effect in new biotechnology that industrial machinery had in modern biotechnology, automating "sequencing factories" where biological data are produced "better, faster, cheaper" (Zweiger, 2001). We can say that intervention in, and manipulation of, natural processes have become the central problem of biotechnology today.

The modes of knowledge production and use have changed accordingly. Classical biotechnology was characterized by trial and error and the refinement of well-known craft techniques. Knowledge was conveyed interpersonally through apprentice-type relationships. In modern biotechnology, rigorous and systematic empirical observation, in the context of controlled experimentation, became the norm. Although certain industrial processes were jealously guarded as trade secrets, most scientists scrupulously recorded and reported their experimental results and observations, making them available for application by industry and colleagues alike. They published their findings in the new journals and presented their work at meetings of scholarly societies. They collaborated in research institutes that brought industrial and academic scientists together around key problems.

In the new biotechnology era, many of these practices have endured. Academic research continues, scientists publish studies, scientific societies still meet, and scientists make presentations. However, controlled experimentation seems to be giving way to synthesis, simulation, and modeling—computational molecular biology—in many quarters. In some respects, this can be seen as a return to the trial-and-error approach of classical biotechnology, but on a scale of industrial proportions. Methodologically, biotechnology researchers are more likely to hunt for genetic sequence patterns, and then try to determine their biological significance, than they are to hypothesize certain structures or processes and then confirm or refute the hypothesis with experimental data. More important, the influence of private-sector firms and funding in academic as well as industrial research has fostered secrecy and competition, rather than public presentation and availability.

Institutional Players and Relationships

Clearly, the modern and new eras of biotechnology have both been characterized by close and productive relationships among academic institutions, scientists, and private firms. To some extent, this is because biotechnology is by definition an applied science situated at the intersection of science and engineering. Although there are important cultural and institutional differences between academic and for-profit research, from the days of brewing and zymotechnology to today's combinatorial libraries and bioinformatics, firms with a commercial interest in biotechnology applications have forged links with university researchers. They have endowed laboratories, projects, individual scientists, and whole departments. Both academic and industry researchers publish in the field's top journals and present their work at the same conferences.

That said, the university–industry alliances in biotechnology have taken a new turn in recent decades with the formation of startup firms that bring together academic talent and private venture capital. Biotechnology today is one of very few fields in which academic scientists commonly hold equity in private firms that commercialize their research while continuing to retain their academic appointments. The 1980s were particularly active years as new changes in intellectual property law encouraged investors and universities alike to experiment with novel organizational forms to support and commercialize research work.

Today the innovative relationships and activities of the 1980s have become a fact of life in biotechnology research. In exchange for academic research funding, private firms expect to retain outright ownership or favorable rights to the results. For their part, many academic scientists have come to consider restraints on publishing or sharing research materials to be a routine and acceptable part of their work (Blumenthal et al., 1997). Universities now engage in essentially the same activities (applied research) for many of the same reasons (revenue generation) that private firms do. Stanford University biochemist and Nobelist Paul Berg observed that universities are just as eager to "capitalize on their intellectual properties" as private firms are because "universities have finally realized that there are potentially lucrative payoffs from faculty discoveries" (Benowitz, 1996, p. 1). They have created new organizational spinoffs whose main purpose is to market inventions. In the current climate, graduate students in biotechnology research programs necessarily learn as much about entrepreneurship, patentability, competitive advantage, and confidentiality agreements as they do about disinterestedness, open communication, acknowledgment and originality, and the ethical conduct of research.

To summarize these themes, we can reflect on the ways that practices, technologies, knowledge, and institutional arrangements work together over time in science. Star and Bowker (2002) argue that these elements should be considered collectively as infrastructure—a combination of material systems and institutional formations that embody and ramify the assumptions, beliefs, and practices of a society. They reiterate the principle set out by sociologist W. I. Thomas: "If men define situations as real, they are real in their consequences" (quoted in Merton, 1948, p. 193). Such a principle seems to have shaped the social and technical infrastructure of biotechnology/bioinformatics as molecular biologists and geneticists came to define genetic matter *as* information (not *analogous to* information) and set the biotechnology research agenda accordingly. They have designed studies and technologies, created organizational and institutional forms, and influenced public policy on the premises that genes are essentially informational and information is a commodity to be generated, circulated, and owned.

THE PROSPECTS FOR SCIENTIFIC COMMUNICATION

On the basis of the preceding discussion, what are the implications for scientific communication in biotechnology? Obviously there has been a long-standing debate about the growing ties between proprietary research and development and academic science because they have traditionally been divergent efforts, the latter to generate profits and thus treating research results as property, and the former to seek knowledge and thus sharing data freely. As mentioned at the beginning of this chapter, there was already concern about the blurring of academic–industrial boundaries in the 1980s. At the time, Zuckerman (1988) notes, "[A] sizeable fraction of university scientists are involved in such collaborations and are thus exposed to new restraints on ownership and communication on a scale previously unknown" (p. 10).

A host of colleagues agreed. Donald Kennedy (1982), then president of Stanford University, argues that biotechnology had ended the separation between basic research performed in universities and the exploitation and marketing of that research by private firms. Kenney (1986) suggests that "The channels of information flow in biology are being adapted to the reality of the market" (p. 125). Philosopher of science Sheldon Krimsky (1988) is more blunt: "[Academic-corporate biotechnology research] linkages have created an entrepreneurial atmosphere that has begun to alter the ethos of science. Norms of behavior within the academic community are being modified to accommodate closer corporate ties" (p. 34).

For 20 years, then, these and other observers have warned against the potential consequences of industry funding and collaboration for norms and practices in the biosciences (see Bull et al., 1982; DeForest et al., 1988; Organization for Economic Cooperation and Development, 1989; Rudolph & McIntire, 1996). Other authors, particularly those personally involved in biotechnology research, conclude that any threats to scientific or academic autonomy can be handled or neutralized by the vigilance of individual academic scientists and more communication and trust among academic–industry research collaborators (Beachy, 1988; MacCordy, 1988; Price, 1988).

Still others argue that the norms of science described by Robert Merton and invoked by critics of academic–industry ties (i.e., values of communality, disinterestedness, organized skepticism, and universalism) constitute something of a straw man. Such ostensible norms have been the subject of a large and enduring body of critique (see e.g., Martin, 1999; Rabinow, 1996). Merton's norms, objectors say, are idealized, and his functionalist analysis of their role in science is naive. Some have suggested that, at best, these norms are balanced by counternorms such as secrecy, commitment, irrationality, and personal judgment (Mulkay, 1976). Mitroff (1974) proposes solitariness, particularism, and organized dogmatism as counternorms.

The debate continues. Yet a number of recent studies, as well as anecdotal reports and case studies, indicate that what was an emerging phenomenon (or looming worry) in the 1980s has become institutionalized and routine today. Although the ethical and policy implications of owning life forms continue to be debated by policymakers, philosophers, and the public (see Best & Kellner, chap. 8; and Murdock, chap. 9, this volume), biotechnology researchers by and large seem to have accepted the market rationale that only ownership rights (patents) will motivate scientists' curiosity and innovation (see e.g., Rudolph & McIntire, 1996; Zweiger, 2001). The growth of knowledge for its own sake, or to improve the human condition, are no longer sufficient motivations for research. Today the dominant motive is the establishment of property rights in information. It has had several important effects on scientific communication as discussed next.

Retreat From Publication

As a consequence of confidentiality or materials sharing agreements, or as a condition of university–industry partnerships, most private research funders now reserve the right to prior review or even the withholding of any draft presentations or publications by their academic scientist collaborators, either to initiate patent applications or preserve competitive advantage. As Robert Rubin, vice provost for research at the University of Miami, says, "It is not unheard of for a company to just sit on an idea, not because they want to develop it but because they don't want anyone else to" (cited in Gibbs, 1996, p. 16). Today the standard practice in biotechnology research is to patent first and then publish. In the end, the results are still public, but they are not freely usable by other scientists as are other published data: "Publication of research results becomes somewhat trivial when the patent filing precedes publication" (Kenney, 1986, p. 124). Ordinarily, publication delays last anywhere from 2 to 6 months, but may be longer.

Although the NIH considers delays of more than 60 days to be "unacceptable" (American Association of University Professors, 2001, p. 69), there are a growing number of cases in which funders have attempted to prevent publication or succeeded in preventing publication indefinitely, either to suppress undesirable findings or maintain the commercial value of a discovery (Gibbs, 1996; Hilts, 2000; King, 1996). An NIH survey of industry–university agreements in 1993 revealed that 22% of the agreements permitted publication delays of more than 6 months (Healy, 1993). Another study showed that 58% of life-science companies sponsoring academic research required investigators to withhold results from publication for more than 6 months (Blumenthal, Causino, et al., 1996). A third survey revealed that about 20% of life-science faculty admitted that they had withheld data from publication for over 6 months to protect the commercial value of the results (Blumenthal, Campbell, et al., 1996). Some observers have suggested that data withholding is more prevalent and normative in genetics than in other specialties (Blumenthal et al., 1997; Weinberg, 1993).

Publication Bias

Recent studies indicate that scientists with industry ties tend to report and publish more findings that are favorable to funders than are unfavorable (Barnes & Bero, 1998; Misakian & Bero, 1998; Stelfox et al., 1998). In a survey of 48 biomedical research journals, less than half (43%) reported having policies about the disclosure of contributors' conflicts of interest (McCrary et al., 2000). However, several prominent journals, including the *British Journal of Medicine* and the *Journal of the American Medical Association*, are considering or have adopted policies requesting that authors declare their industry ties or relevant interests (*Economist*, 2001a). The U.S. Public Health Service and the National Science Foundation require that research institutions funded by the agencies report investigators' industry ties. However, the institutions show "substantial variation" in policy and implementation (McCrary et al., 2000).

The scientific literature can also be biased by what is or is not published. This can be a consequence of the retreat from publication. Yet unpublished materials such as student theses and dissertations are also affected. Several academic–industry research agreements have stipulated that student theses or dissertations based on industry-funded research are subject to prior review by funders. In some cases, dissertations have been deemed proprietary information by funders and been sequestered temporarily or withheld entirely (Kenney, 1986; Krimsky, 1988).

The organization and access to published and unpublished materials also biases the literature for those searching it. For example, although the internet has become an essential resource for data, unpublished research reports, and informal interaction, as well as published materials, some observers have argued that the information online has become ever more narrowly specialized and "Balkanized" (Van Alstyne & Brynjolfsson, 1996), or that it tends to lead searchers to only the most popular or hyperlinked materials (Glanz, 2001).

Erosion of Peer Review

Several of the studies cited earlier suggest that the prevalence of industry ties among academic researchers may make it difficult for journals to find enough disinterested reviewers for some manuscripts. In fact the *Los An-*

geles Times revealed recently that 19 out of 40 drug review articles published in the "Drug Therapy" section of the *New England Journal of Medicine* over the prior 3 years were written by scientists with financial ties to pharmaceutical companies. After an inquiry, it was found that the authors had indeed disclosed their drug industry ties to the journal, but their reviews were published anyway on the grounds that it was difficult to find reviewer/authors without industry ties (Warde, 2001).

Occasionally, scientists report delaying or withholding research results because of suspicions that in an increasingly competitive, market-oriented research environment, reviewers might steal ideas and jeopardize their priority claims (*Economist*, 2001a). Another factor in the erosion of peer review has been the release of research results in the popular media before they have been peer reviewed to stimulate investor interest. Perhaps the most prominent recent example was the joint announcement of the complete sequence for the human genome by Celera and the Human Genome Project in June 2000 (*Economist*, 2000; *The New York Times*, 2000). Despite the extensive coverage, however, the announcement was later characterized as somewhat anticlimactic because "there were really two genome surveys and each was still short of its respective goal" (Zweiger, 2001, pp. 204–205). The Human Genome Project group had planned to announce a draft of 90% of the genome using a database of samples from hundreds or perhaps thousands of people, but at the time of the announcement had only sequenced about 85%. In contrast, Celera had set as its goal 100% sequences of five or more people repeated 10 times each to ensure reliability, but at the time of the announcement it had sequenced only one person's DNA an average of less than five times (Zweiger, 2001). Moreover, whereas the Human Genome Project made its data immediately available to anyone, Celera released its data only to carefully selected colleagues.

Constraints on Informal Interaction

Informal contacts among scientists are also influenced by academic–industry ties. Scientists and students have reported that confidentiality agreements or competitive pressures have compelled them to restrict their informal interaction with other researchers (Gibbs, 1996; Kenney, 1986). As Paul Berg notes, "Certainly [secrecy is] changing the way science is done from the way it was done 10 or 15 years ago" (Benowitz, 1996, p. 1).

Blumenthal et al. (1997) found that only about 9% of the 4,000 academic life-scientists they surveyed reported that they had refused to share research results or materials with other university scientists in the previous 3 years. However, 34% of their respondents said that they had been denied such results or materials when they asked other university scientists for them. The investigators found that genetics researchers "were significantly more likely than other life scientists to report having refused other faculty access to research results" (p. 1226). Scientist respondents involved in the commercialization of their research, genetics researchers, and those with higher publication rates were more likely than other respondents to report refusal to share research results.

CONCLUSION

Consistent with the changing institutional and knowledge environment in the field, scientific communication practices in biotechnology have shifted significantly over the last 20 to 30 years. Looking at the broad institutional history of biotechnology reveals that the nature and products of biotechnical knowledge have gone from applied craft skills and experience, to hypothetico-deductive knowledge, to inductive pattern finding. The modes of knowledge production and use have evolved from trial and error, to controlled experimentation and hypothesis testing, to computational modeling and simulation (which some might say is just a mechanized return to trial and error). Institutional players and relations in the field have likewise expanded beyond the guilds and trades of classical biotechnology, to the industries, private laboratories, colleges, and research universities that arose in the modern period, to new biotechnology's academic–industry alliances centered on "faculty entrepreneurs." All of these changes have helped shape, and have themselves been shaped by, a growing cultural, economic, and policy conflation of natural phenomena (especially genetic material in this case) with concepts of information and property. This sociotechnical environment has fostered corresponding changes in scientific communication, away from the traditional practices of scientific research established in the modern era and toward practices reflecting the commercial and competitive pressures of private-sector business.

In summary, at least in biotechnology research today, Nelkin's normative question ("Who should control scientific information?") has been replaced by Zuckerman's contractual/legal question ("Who has rights of ownership in science?"). The answer to the latter is private firms and academic institutions that treat scientific knowledge as an informational commodity to be traded in the market. The evidence suggests that scientific communication in biotechnology conforms more to Mulkay's or Mitroff's counternorms (or, put differently, to market demands) than to Merton's norms of science. We are faced with a system of scientific information and communication that is increasingly based on secrecy/solitariness, commitment/particularism, irrationality/organized dogmatism, and personal judgment and interest. This is the case despite the fact that when asked most scientists ascribe to and affirm the traditional norms as part of their training and practice. Violations of the norms may occur, but are considered occasions for embarrassment (Benowitz, 1996; Rosenberg, 1996). For example, in a recent study of life scientists' withholding of data or research materials, Blumenthal et al. (1997) speculate that their respondents may have underreported reasons for withholding that would have seemed contrary to the norms (such as personal or university financial interest, contractual obligations, or to preserve a scientific lead) and overreported reasons that are more socially acceptable (overwork).

Obviously, scientists must and do continue to share their findings, at least with selected colleagues, or their inquiries cannot be sustained. Yet if biotechnology is a guide, the information flows are increasingly private, informal, and specialized (i.e., more parochial and less universalistic). Reputations are still critical in the scientific reward system of biotechnology. Yet they seem to be built more on insider information, shared within closed teams of collaborators or networks of institutional alliances, than on freely circulated results that anyone else may review, reproduce, or use. Rapid dissemination of provisional findings to a carefully selected list of colleagues (the familiar process of "trusted assessorship") may be considered a more useful and generative step than conventional peer-reviewed publishing. Although scholarly publishing continues to offer some credibility within the discipline, if visibility is needed, then popular media are much more powerful in terms of attracting funding, political allies, administrative resources, talented collaborators, and student workers. Overall, the situation lends itself better to highly selective, specialized media like the internet or videoconferences, or to highly controlled outlets like press conferences, rather than peer-reviewed publications, conference proceedings, or even teaching. In short, new-era scientific communication and information management look a great deal like private sector communications and image management—competitive, committed, closed, strategic, and risk averse. Information, interactions, and impressions must all be managed carefully to achieve the greatest effect and benefit and to avoid loss.

In conclusion, we can say that biotechnology may have demonstrated that the counternorms can provide a cohesive and powerful basis for a new kind of science and for new institutional forms where it can be practiced. The question remains, however, whether biotechnology will remain an outlier or become a leading indicator for other academic disciplines.

ENDNOTES

1. Public access versus professional control, rights of access versus obligations of confidentiality, competitive secrecy versus open communication, and national security versus scientific freedom.

2. A great deal of the historical overview in this section is drawn from the excellent history of biotechnology by Bud (1993), as well as other sources.

3. A more recent example is a 5-year, $25 million agreement between UC Berkeley and Novartis, in which the Swiss firm has licensing rights to a proportion of discoveries made in Berkeley's Department of Plant and Microbial Biology, "equal to the company's share of the Department's total research budget, whether or not the discoveries result directly from company-sponsored research" (American Association of University Professors, 2001, p. 69).

4. Zweiger does not speculate on the epistemological significance of this shift for science. It might be said that social scientists have employed such techniques for decades owing to the complexity of social phenomena they study. Yet in doing so, they have been criticized as immature sciences at best, or pseudosciences at worst, to the extent that they have departed from controlled experimentation and hypothesis testing.

INFORMATION AND POWER

Transborder Information, Local Resistance, and the Spiral of Silence: Biotechnology and Public Opinion in the United States[1]

Susanna Hornig Priest
Texas A&M University

Toby Ten Eyck
Michigan State University

Medical and agricultural biotechnologies are being touted by some as the answers to such problems as world hunger and disease.[2] The conventional wisdom with respect to these new technologies is that people within the United States are at least neutral and for the most part extremely positive about them, whereas populations in Europe and parts of the developing world, such as India and Africa, are very pessimistic. Explanations offered within the scientific community for the differences in opinion, perhaps not surprisingly, often center on education, suggesting that if only people understood these technologies more fully they would appreciate them and not be "afraid" (Garland, 1999; Priest et al., in press).

This "ignorance and superstition" explanation fits more neatly with our perceptions of the developing world, however, than of Europe. For an explanation of public opinion concerning biotechnology in Europe, one alternative explanation has been the effects of the media, particularly the tabloid press. In the UK, at least, the tabloids have undergone relatively recent expansion in comparison with the older prestige press with its tradition of restraint in the context of public controversy. Tabloids have subsequently become a scapegoat for all kinds of public relations problems—from mad cow hysteria to the death of Princess Di. The tabloids' use of the "Frankenfood" designation and similar representations have been widely blamed for turning Europeans against biotechnology (Gaskell et al., 2001).

Other explanations of European public opinion are explored by Murdock (chap. 9, this volume).

Of course this is *magic bullet* thinking about uniform, direct, and short-term media effects that is insupportable on the basis of what is known today about the relationships between media and public opinion. There is a need, therefore, to dig deeper. The first place to look is at the public opinion data. Media representations are often confused with public opinion. Although the two are indeed intertwined—and although media representations of public opinion often stand in for the actualities as inputs into policymaking—they are not necessarily equivalent. U.S. public opinion regarding biotechnology as measured by opinion polls, as well as a variety of indirect measures such as letters to the editors of major newspapers, shows considerably more variation than many media accounts acknowledge. Nearly as many people in the United States believe that genetic engineering will make the quality of life worse as believe this about nuclear power (Priest, 2000), yet few observers argue that nuclear power is noncontroversial in this country. The U.S. Department of Agriculture (USDA) was overwhelmed with protests about its recent plan to allow biotech foods to be labeled "organic." Several major U.S. food manufacturers have disavowed the use of GMO ingredients—an action presumably taken on the basis of market considerations. Biotechnology is controversial in the United States, and it has become gradually more so over the past 20 years (Priest, 2000, 2001a). Those in the United States may be somewhat more positive, on average, than populations in at least some parts of Europe, but we are not monolithically in favor of these technologies. In addition, objections are not confined to a Luddite sliver of the population, but include people from higher socioeconomic classes and those with a college education who view other technologies as progressive and advantageous. Despite the fact that a majority of the U.S. population does lean toward favoring these technologies, opinion appears quite polarized (Priest, 2000).

Further, both within the United States and for Europe, the science literacy hypothesis is inadequate as an explanation of variations in public attitudes (Priest, 2001a, 2001b, 2001c; Priest et al., in press). On the basis of recent national survey data, trust in the major individuals and institutions that are bringing us biotechnology—scientists, farmers, grocers, and industry—is more important than levels of knowledge in predicting support for various applications of biotech. These latter influences exist, but are quite modest, as well as being difficult to decouple from interest in and general support for science. Discomfort appears to be creeping up among the most highly educated segment of the U.S. population (National Science Board, 2000).

The real question, therefore, is not why people in the United States do not raise questions about biotechnology, but why the U.S. press has in general failed to reflect the questions that are in fact regularly raised. It is a question worth asking because the pattern of media coverage on biotechnology-related issues in the United States has permitted policymakers to assume there is no popular resistance to these developments. In turn, the failure of voices of dissent to find an expression in the press has arguably fueled the escalation of protest, becoming at times violent.[3]

These concerns with biotechnology seem to revolve around the response of the public, and policymakers, to risk. Risk is a basic factor in developing our sense of self and identity; most seek connections that are stable and risk averse (Giddens, 1991). Life is characterized by navigating through contingencies in which stable relations are paramount. However, risk is socially constructed, and the architecture of risk becomes a defining principle as we seek to avoid hazardous circumstances (Beck, 1999). According to Slovic (1987), a major component of this architecture is the perception of whether the risk is voluntary. If risk is perceived as voluntary, an individual is much more likely to view it as acceptable than if it is not. Biotechnology, especially in its agricultural applications, is viewed as involuntary and therefore unacceptable. We can choose whether to jump out of an airplane with a parachute, but we cannot control what goes into our food on the farm or on its way to us.

Ritchie (chap. 2, this volume) looks more closely at some confusions derived from framing biotechnology issues in terms derived from information science. For our interests, Hall's (1980) idea of the media as a "site of struggle" between different interpretations or readings of events moves us toward thinking about various decodings among audience members. Hall assumed, as our empirical data suggest, that powerful institutions acting as sources for media material (in this case, primarily industry and science, but potentially also other relevant institutions of government, commerce, agriculture, etc.) create the frames through which audiences and readers interpret content. However, unlike deterministic theories of framing effects, Hall's theory assumes that those audiences actively impose meaning on the messages they receive, rather than react passively to accept the frame with which they are presented. In other ways, Hall proposes that audiences decode or interpret messages in ways that are not always intended by their creators. The result may be what Hall calls *oppositional* readings.

ALTERNATIVE EXPLANATIONS

Alternative explanations are available both for the nature of media coverage and the existence of alternative readings of the media. As this chapter tries to demonstrate, the mass media can be a force for the suppression of dissent, but they are also a forum for its expression. For biotechnology, resistance may initially emerge at the local or regional level. Institutional interests routinely seek to offer information subsidies that help frame problems in ways they see as advantageous to their positions, and biotechnology is no exception (Plein, 1991). Thus, the differences between local and national news frames, to the extent they are systematic, are nontrivial. Our observations problematize the differences between national and local news frames in U.S. coverage of controversy.

The Framing of Biotechnology

Several theories regarding influences on media content are useful in understanding the reluctance of the U.S. media to accurately report on public concerns regarding biotechnology research and applications. Framing, or article emphasis, is widely believed to suggest a particular context for interpretation to readers or audiences. The term *framing* was introduced by social psychologists, political scientists, communication researchers, and others as an experimental variable—a message characteristic that can be manipulated. The empirical evidence is mixed; when the concept is turned into an independent experimental variable by social or behavioral scientists, measurable effects may vary. Undoubtedly, too little is known about the conditions under which framing may be influential. Yet a part of the problem is likely to be that in the interests of creating controlled conditions the experimental results ignore that framing and frame interpretations are social processes that take place in certain political and social contexts that are broader and more complex than either experimental settings or isolated news stories. Although both experimental results and the more general concept are useful in understanding public opinion formation, relevant factors extend beyond specific message characteristics and characteristics of individual receivers to the political and social conditions.

Gandy's (1982) concept of information subsidy as it applies to technical controversies provides some insight into ways in which the media have been heavily influenced by university and industry research perspectives. This involves those with a vested interest in the technologies feeding reporters specific information—information subsidies—which casts these technologies in a positive light (Priest & Talbert, 1994; Ten Eyck et al., 2001). The motivation of those involved in biotechnology research to provide such information subsidies has surely increased along with the strengthening of their ability to assert intellectual property rights over their findings and, thus, to reap potentially enormous economic benefits from them (see Lievrouw, chap. 6, and May, chap. 5, this volume).

Noelle-Neumann's (1984) theory of spiral of silence is helpful suggesting that some potential voices are silenced by the media, either through neglect or belief that opposing points of view are not important. Once this spiral has begun, either on the part of reporters or opponents, these views are further silenced. Hallin's (1989) idea of media legitimization processes adds another piece to the puzzle by looking at how the media both legitimize specific spokespersons and become vehicles for dissemination of legitimized information. Given that most people get their information concerning biotechnology from the mass media, those who are quoted in these channels and labeled as experts come to be seen as such by the public. This is true even if the public disagrees with the message—a scientific expert is still a scientific expert even if I disagree with his or her statements. In addition, because most people are not conducting genetic engineering experiments in their basements, they must rely on the media for information concerning work being done in this area.

Studies that note the discomfort of journalists with scientific uncertainty (Friedman, Dunwoody, & Rogers, 1999) and the long-standing tendency in the U.S. press to promote science and technology as inherently progressive (Mulkay, 1997; Nelkin, 1995) are also useful. In addition, the perspectives of powerful U.S. institutions are generally overrepresented partially as a result of information subsidies. Other factors leading to an overreliance on official sources include lack of local competition, dominance of the Associated Press (AP) wire service, and increasing economic concentration in the media industry. This reliance on official views is evident in coverage of biotechnology because the field is both large-institution science and industrial science.

Oppositional Readings

The existence of oppositional readings is what makes it possible for the media to serve as sites of struggle (Hall, 1980). This notion provides some insight into those isolated cases where the existence of opposition has become suddenly visible against a backdrop of subsidized media legitimizing only mainstream (probiotechnology) opinion. The cloning controversy—in which those who believed cloning (even of humans) should be permitted to further science and pursue medical goals were opposed by those who feared potential medical, environmental, and ethical consequences—provided such an example. In this case, however, the increased visibility and legitimacy afforded ethicists' perspectives did not seem to have a significant short-term impact on either regulatory policy or public opinion (Priest, 2001b). Although media agenda-setting effects are well documented, framing may or may not be influential in the cases of emerging controversies such as biotechnology (Priest, 1995). Some authors (e.g., Leahy & Mazur, 1980) have speculated that media coverage of concerns about technological risk always has a negative impact on public opinion. However, although Nisbet and Lewenstein (2000) predicted a negative turn in public opinion will follow from the increased attention to ethical dimensions for biotechnology generally, the changes in U.S. public opinion have been gradual and seem as likely to have predated media attention to these issues as to have resulted from them. In other words, although media stories are an obvious source of information on the issues for the nonscientific public and may not be without effect, particularly as an agenda setter, the assertion that media fully determine public opinion in this area does not seem supported. National media coverage may well have followed, rather than led, the emergence of public concerns.

Biotechnology also challenges important American values—from the notion of a genetic heritage to environmental stability to economic independence. The involuntary nature of biotechnology has been reinforced by some highly publicized breakthrough events that also highlight failures of the regulatory system. One such event involved Starlink corn, which despite having been approved only for use in animal feed was found by environmental watchdog groups to be present in food products destined for human consumption. Another breakthrough event occurred when maverick scientists in Rome announced their intention to clone people, prompting scientists working on the cloning of various agricultural animals to express dismay in the light of significant but heretofore unpublicized challenges and failures in their own work. A third was the discovery that modified corn genes had been spread across much of Mexico. Proximity to an event makes it more likely that it will be seen as a breakthrough because concern about effects will be keener; thus, local media coverage of concerns about local experimentation with biotechnology generates issues that sometimes rise into awareness of the larger public and additionally serve as breakthrough events. Survey data show broad public concern in the United States, as elsewhere, over the adequacy of the regulatory system to deal with these issues and to incorporate moral considerations into decision making regarding their acceptability.[4] Thus, these occurrences serve as lightning rods for growing but not fully articulated concerns about the adequacy of the oversight system currently in place and the latent problem of forcing involuntary risks on a population.

There has been a suggestion of a transborder spillover of European concerns to the U.S. population. If there were indeed a spiral of silence in the United States until recently, then the impact of European dissent as expressed in restrictions on international trade in agricultural commodities may also have contributed to breaking that spiral. Although this particular hypothesis is difficult to test or evaluate in a formal way, it is likely that public discussion in the United States in response to the European situation served to make concerns regarding genetically modified organisms (GMOs) more visible as well as to legitimize dissenting views on their acceptability.

Finally, it is also possible that the greater political diversity characteristic of the press in much of Europe compared with that in the United States might have stimulated debate there at an earlier stage, while the U.S. press remained comfortably mainstreamed. Wilkie and Graham (1998), for example, suggest that U.S. media coverage of the cloned sheep Dolly included more voices from the scientific community than did the parallel press debate in the UK.

Theories that suggest factors suppressing coverage of dissent against biotechnology in the media can also be used to understand the growing reflection of a diversity of opinions. Biotechnology is an industrial enterprise, and biotech companies are among those powerful institutions in the United States offering information subsidies to mass media outlets in the forms of press releases, "public service announcements" from biotechnology councils, and advertising. Some even assert these organizations are intertwined with large media corporations (Kerbo, 1991). With the rise of consumer activism in the United States, however, led by such visible individuals as Ralph Nader and Jeremy Rifkin, groups opposing biotechnology have become strong enough to offer the same kind of information subsidies. Reporters can no longer ignore these alternative perspectives without paying some price.

Similarly, although the concept of the spiral of silence offers some insight into the decades of failure by the media to cover dissent against biotechnology, it can also help us understand why turns in both media coverage and public opinion have appeared so quickly. On occasion, U.S. mass media have reclaimed their role as sites of public struggle among opposing interests; between such incidents, dissent (while visible in the polls) is considered an individual opinion and remains effectively concealed from public arenas. The shock to the U.S. system generated by European resistance to GMO foods—and its impact on the ability of the United States to export foods—along with a series of breakthrough events have provided focal moments for attention to the kinds of strong voices that serve to launch or redirect patterns of both media coverage and public thought.

The remainder of this chapter uses public opinion data to explore (a) the extent and ways in which breakthrough events and spillover from European public discourse contributed to the legitimization of dissenting views in the U.S. press *that were already present* as a significant component of public opinion; (b) the extent to which Hall's notion of the media as a site of struggle between dominant and oppositional messages is useful as an explanation of these dynamics; and (c) implications of the case of biotechnology for further research into ways the national U.S. press may limit and constrain the expression of dissent in cases of technological controversy.

PRESS COVERAGE AND PUBLIC OPINION

This analysis rests on a comparison between data on U.S. public opinion on biotechnology-related issues (unpublished 1997 data collected by Jon Miller and others[5] and more recent follow-up survey data based on similar questions collected by Priest in the spring of 2000) and content analyses of media coverage of biotechnology (one study covering the period of 1992–1996, prior to the first opinion survey, and one covering the period 1997–1999, prior to the second opinion survey; Ten Eyck et al., 2001).

The public opinion data reveal at most some moderate elevation between 1997 and 2000 in the percentage of U.S. respondents who felt that genetic engineering would have a negative impact on quality of life. Further, comparisons with data collected in 1982 and 1986 by the U.S. Office of Technology Assessment set these data in the context of a more gradual trend because negative attitudes toward biotechnology were also present at that time, although to a lesser degree. These data suggest that serious concerns about the potential risks of biotechnology in the U.S. population are long standing. In fact the conclusion from the 1987 Office of Technology Assessment study stated that a majority of Americans believes strict regulation of biotechnology is necessary (Hamstra, 1998). At the same time, this same study found that a majority of respondents did favor genetic engineering when it would be useful to resolve a variety of medical and environment problems. Age, education, and level of interest in science had no apparent effect on these attitudes. Although a large majority of the respondents approved of medical applications (75%–86% on the various applications), nearly three quarters of respondents were concerned with the morality of human cell research. This overlap shows that many of the respondents viewed genetic engineering on a case-by-case basis, agreeing with one statement while disagreeing with another.

Opinion surveys in the 1990s showed similar results. Hoban et al. (1992) find that opposition to genetic engineering was widespread, with moral objections driving much of that opposition. The Biotechnology Industry Organization argue that its focus group study in 1994 found little public knowledge of biotechnology, although focus group participants did show concern over some of the possible products produced by biotechnology (Hamstra, 1998). It should be noted that the former research was based on surveys, whereas the latter was based on focus groups and conducted by a probiotechnology group. Still, that attitudes could be manipulated to this extent reflects that both the promises and risks of biotechnology are plausible and opinion on this technology is currently in its nascent stage.

These findings, coupled with articles from *The New York Times* and *The Washington Post* in the mid-1970s, which discuss scientific and public opposition to genetic engineering, highlight that opposition to these issues has been present from the start, and recent findings do not reveal a new social phenomenon. On July 18, 1977, *The Washington Post* ran an article that discussed the pros and cons of biotechnology, reporting that although many scientists felt that the dangers associated with genetic engineering had been exaggerated, others were still concerned that research in the area was moving too quickly and still very dangerous (Cohn, 1977). On June 28, 1978, *The New York Times* published a letter accusing the National Institute of Health (NIH) of bypassing genetic engineering safety regulations (Simring, 1978). Although these articles and others like them are available, they are not necessarily the norm. Still, the long-standing presence of opposition is what might be expected from examination of public response to other controversial food and medical technologies in the past, such as pasteurization (Ten Eyck, 2001), vaccinations (Baker, 2000), and the abortion pill RU486[6] (Clarke & Montini, 1993).

THE PROBLEM

The statistical relationships between shifts in media coverage and public opinion trends suggest it is unlikely that national media coverage in the United States bears the responsibility for changes in the way public opinion regarding biotechnology has evolved. According to results we have published elsewhere (Ten Eyck et al., 2001[7]), elite press coverage (*New York Times* and *Washington Post*) from 1992 to 1996 was 42.5% positive and 4.4% negative; equivalent coverage from 1997 to 1999 was 31.1% positive and 6.0% negative. (A large increase occurred in the "no evaluation" category, which went from 42.0% between 1992 and 1996 to 52.8% in the second wave.) Despite this roughly 10% decrease in positive media treatment and a parallel increase in more balanced coverage, only a tiny change was noted in the public opinion data for 1997 and 2000, which moves only from 50.5% positive and 29.1% negative to 52.8% positive and 30.1% negative for projections of the impact of genetic engineering on quality of life. Note that both positive as well as negative opinion percentages increased, albeit not dramatically; no overall trend during this period other than this small increase in polarization is easily discernible. Note also that the prevalence of negative public opinion is much greater than the percentage of negative media opinion in periods leading up to the surveys. Part of the gap between press and media coverage may be explained by the idea that negative press or communication carries more weight than positive press or communication (Sapp & Harrod, 1990), and that it takes only a few negative stories to shift public opinion. In addition, trust in government agencies has been declining over the years (Chanley et al., 2000), and governmental sources have a high degree of visibility in biotechnology stories (Ten Eyck et al., 2001). Although these are plausible explanations, we feel we must look elsewhere for the main sources of negative public opinion.

According to the same study by Ten Eyck et al. (2001), biotechnology was framed as *progress* in 48.9% of the 1992 to 1996 articles, but 62.9% of the 1997 to 1999 articles. Although the frame of the articles included in this study was coded independently of positive, negative, or neutral tone, discussion of "progress" clearly invites a positive way of thinking about science or technology compared with ethical discussions or the idea of runaway technology. Despite the move toward more balanced coverage in the sense of coverage that is less fully dominated by positive, probiotech material, there has obviously been no erosion of the tendency to equate biotechnology with progress. Far from being abandoned during the past decade, this emphasis was further embraced.

Finally, the three frames that seem, conversely, to imply or entail inherently critical perspectives or problems—those that we classified as the "Ethical," "Pandora's Box," and "Runaway" frames—together accounted for 10.4% of the earlier set of articles, but 16.0% of the later set. Independent of the articles' tone becoming more neutral and regardless of the increasing focus on the progressive nature of the research, an increase in discussion of possible problems or concerns associated with biotechnology was indicated.[8] Of three different indicators of the tone of coverage—the positive–neutral–negative dimension, the presence or absence of the progress frame, and the presence or absence of frames problematizing biotechnology—two indicate more diversity of opinion in U.S. news media representations in the more recent period. Yet this change appears to follow, not lead, the change in public opinion that evidently occurred between 1986 and 1997 and now appears to have largely stabilized.

Although it is possible that changes in media coverage occurring earlier than 1992 helped influence later changes in public opinion, on the basis of the data available, the "Occam's razor" (simplest) solution supports the "spiral of silence" interpretation rather than a more direct media effects hypothesis. Public opinion in the United States appeared to change before—or at most simultaneously with—media content, which included both news articles and opinion pieces. However, this shift remained largely invisible in national mainstream media accounts and the national policy dialogue, in which positive public opinion was generally assumed, until much later. This is not a new phenomenon; foreign policy is often based on perceived public opinion. As Kull and Ramsay (2000) observe, "once a belief about the public becomes established, there is no reliable corrective mechanism" (p. 109). If this is the case with biotechnology, the absence of some kind of national outcry may have led policymakers to believe the public was accepting of this new technology.

Framing

Public opinion polls serve, among other purposes, as tools of social control designed to help achieve the goal of "manufacturing consent" (Herman & Chomsky, 1988) by providing feedback to elites on their success in this endeavor—and possible trouble spots. Biotechnology undoubtedly provides a case in point. *Resistance* is defined as "the problem" and *science education* as "the solution." Defining opposition as ignorance enlists new actors—including both educators and science journalists—in the battle to contain and defeat it. As we hope has been demonstrated in this study, poll data can also be used to both document the existence of dissent and suggest alternative interpretations and explanations for its relation to media content. Understanding dissent more fully requires a different sort of theory, however, and here Hall's (1980) notions of the media as a site of conflict can play an important role.

Political and social circumstances of importance to both framing and interpretation of media content regarding biotechnology have already been suggested. In the example of cloning, such profound ethical questions were raised for many observers and writers—and, we presume, so many audience members, although the evidence is not direct—that dissent "broke through" into mainstream U.S. media accounts. The dissent reflected both concerns of the religious right as well as the environmentalist left, so that hegemonic processes may have been inadequate to contain it. Further, institutional actors publicizing the Roslyn Institute's success with the cloning of an adult mammal, media organizations seeking remarkable stories, and scientific publications seeking prominence for their research reports all contributed to focusing attention on a Scottish sheep (Priest, 2001b; Wilkie & Graham, 1998).[9] The furor over cloning resulted from a constellation of institutional, not simply individual, factors. However, academic and religious ethics experts were brought in to repair the perception that matters were in disarray (Thompson, 1997), although the scientific displacing of popular wisdom resulted in little long-term change in regulatory policy beyond the preexisting U.S. ban on federal funding for fetal cell research (Bettelheim, 2001). Meanwhile, other issues—ranging from concerns over genetic engineering of everyday supermarket foods to the issue of indigenous people's intellectual property rights over local genetic resources to the economic impacts of genetic research on farmers and consumers—received little attention.

GM foods are also under attack in the United States to some degree, although the U.S. reception has been somewhat more positive overall than the reception in Europe. The U.S. FDA has been reconsidering both its labeling policy and its policy on mandatory review of bioengineered food products—steps that reflect de facto recognition of public concerns, which is a different approach than that taken on cloning. It certainly cannot be argued that suppression of public debate or dissent in the United States about GMOs was entirely successful longer term. The impact of local biotechnology issues and local media coverage of them, understood through Hall's lens on oppositional framing, makes it possible to see local as well as European news as sources for information on breakthrough events capable of shifting both public opinion and media reportage of it.

European Discourse as Breakthrough Event

Public reactions and events in Europe suggestive of widespread resistance to the products of recombinant DNA technology could have influenced U.S. public opinion, but it is hard to see how such an influence would have occurred outside of the popular national media. This is not to say that all nationally dispersed media sources were treating biotechnology in a positive way because environmental and organic magazines were critical of the technology (e.g., Berle, 1988). The European Union appeared as an actor in only 0.1% of the 1992 to 1996 articles and 1.1% of the 1997 to 1999 articles, with other international players appearing in only 0.3% and 0.4% of them, respectively. Transborder information flows reflecting European and other non-U.S. dissent may well have helped legitimize dissent in the United States, a nontrivial effect, thus breaking the spiral. Yet international dimensions did not figure prominently in U.S. coverage and were unlikely to have been a root cause of U.S. resistance, which appeared and grew—largely despite public near invisibility at the national level—in earlier years. For this reason, we must begin to consider other information sources.

We do not want to overstate the case for the indigenous origins of most U.S. opposition. Our content data, based on a sample of elite print sources, reflect only a small portion of the national news. They do not favor front-page coverage over news printed elsewhere in the paper, or between headlines (which were not coded) that entice the reader with compelling images and those that bury news of European dissent within discussions of trade relations, genetic research, or other technical stories some readers will not pursue. We did not take into account TV news, likely the most compelling source in the United States of powerful images of protest activities in Europe. However, we do believe it is fair to conclude that the national media discussion of biotechnology was not weighted heavily toward coverage of European protests, especially during the years U.S. public opinion was forming—quietly—on these issues. Again we do not mean to overstate lack of negative discourse at the national level. Individuals such as Jeremy Rifkin and Ralph Nader were being heard in the press as well as opposing viewpoints being displayed in opinion columns. Still the overall impression from our sample of 1,600 articles is that the press was much more likely to give biotechnology issues a positive slant.

Further, the influence of even a few stories about European reactions on journalists covering issues of science, technology, agriculture, health, and international trade, who would have been more attentive to even rare mention of non-U.S. events and opinion, was likely greater than our statistics would suggest. In other words, although published mention of European perspectives in the sources we studied was quite rare, this does not mean that U.S. awareness of European dissent was not important or did not influence U.S. media in less direct but no less substantive ways. Contributing in a relatively narrow or limited way to the redefinition, legitimization, and highlighting of objections and concerns in the United States, for example, could be an impact of significant proportions on the climate of opinion, especially in elite circles. However, it is still difficult to propose news coverage of European reactions as a major explanation of growing U.S. unease among members of the general public, although it might help in understanding the emergence of more negative press coverage.

In summary, although the direct influence of opinion from outside the United States was probably small, its influence on elite thinking—the thinking of those opinion leaders with the most power over policymaking—was probably greater. At a minimum, it helped direct attention to the existence of dissent within the United States, as well as possible concerns with international trade. This is true even though some biotechnology is undoubtedly here to stay and U.S. policies for GMO foods remain far less restrictive than policies in many European countries.

Local Breakthrough Events

If public opinion in the United States has—for quite a long period of time—been more diverse than the national news agenda has suggested, we must look beyond that agenda for insight into how public opinion for these issues has been formed. Most studies of U.S. news content, including those of science-related news content, focus on the elite press—*The New York Times*, *The Washington Post*, and a small handful of other *prestige* publications such as AP (Evans & Priest, 1995). One justification often given for such choices has been the powerful role of these publications in setting the national news agenda and thus influencing the national policy agenda. Although this explains one mechanism through which dissenting opinions for biotechnology, as for other issues, can be suppressed and helps provide insight into why U.S. opinion was viewed, until recently, as monolithically probiotechnology, it also suggests we must look elsewhere for the seeds of change vis-à-vis public opinion in this case.

Many controversies involving biotechnology—from the 1970s through the present—have arisen at the local level and have been most readily visible in local news accounts, only much later, if at all, reaching the attention of the national press. Although the typical local paper in the United States is a member of the Associated Press and most stories published locally would thus be available for AP distribution, a local story must be seen as having national significance to become nationally prominent. Given the top–down structure of the AP, in which news taking place in large urban centers (e.g., New York; Washington, DC; Los Angeles) usually takes precedence over news from less populated areas (Tuchman, 1978), it is not surprising that local stories from outside New York and Washington, DC were not picked up by newspapers in these cities. Of the 1,374 articles we have coded from *The New York Times* and *The Washington Post*, only 125 (9.1%) were wire service articles (although we do not know how many of these stories were then sent to local newspapers affiliated with these newspapers). News editors elsewhere act as a gatekeeping network, selecting only a few stories with particular news values from many candidates, and often seek local experts to comment on national stories (Ten Eyck, 1999). Agenda setting from one part of the press to another can, however, take place in a bottom–up direction rather than the top–down one conventionally assumed. Local struggles may involve themes, frames, and actors that are distinctly different from those prominent at the national level. Protests related to biotechnology research or other activities are in fact quite regularly reported at the local level in the United States, which we discuss later. However, their impact on the national news agenda and national-level policy discussions is often limited.

Perhaps the original instance of local dissent—or at least the most routinely cited early case—arose from the objections of the Cambridge, Massachusetts, community to recombinant DNA research being conducted by universities there in the 1970s. This resistance resulted in a local moratorium on the research in 1976 and the subsequent formation of the Cambridge Experimentation Review Board, a local citizen review panel that adopted a *citizen jury* model for weighing expert testimony on both sides. The Cambridge case is widely cited as a successful example of citizen participation in science policy formation at the local level (Waddell, 1990), perhaps because the ultimate outcome was the lifting of the ban, widely seen as a triumph of the forces of light over those of darkness. The board was directed to focus on threats to local public health, although less tangible concerns inevitably arose in the course of its deliberations. Yet the case seemed to have attracted relatively little national attention at the time. According to a Lexis–Nexis search, *The New York Times* ran seven articles on the topic between June 17, 1976, and February 8, 1977, when the Cambridge City Council voted to lift the moratorium on genetic research. The *Wall*

Street Journal also published one article on February 9, 1977, to report the City Council vote. None of the articles made it to the front page, with a page 8 article being the closest.

Bovine somatotropin, or bovine growth hormone, a Monsanto product used to stimulate milk production in dairy cows, stimulated further resistance to biotechnology at the local level after it was approved for farm use in the United States in 1993. This was the first widely publicized commercial use of biotechnology in food production.[10] Objections to the product in small-farm dairy states such as Wisconsin and Vermont were widespread; these originally arose on economic grounds because of projections that the product would advantage larger farms over smaller ones. Industry objections again focused the debate on public health concerns, however, whether for strategic purposes or as a result of a simple misreading of the original objections (Hornig, 1991). Objections based on the assertion that the drug would increase mastitis—a disease of the udder—among cows, scare consumers, and threaten both the family farm and the federal milk price support system also surfaced in AP stories printed in the Madison, Wisconsin, *Capital Times* (Greene, 1993a, 1993b), but did not seem prominent in the national press. Again a Lexis–Nexis search shows 10 articles in *The New York Times* that discussed both bovine growth hormone and Wisconsin, although two of these articles were editorials, and one article in the *Wall Street Journal* between 1990 and 1995. One article published by *The New York Times* on February 4, 1994, did appear on the front page. Meanwhile the *Wisconsin State Journal* ran an editorial by the president of the Wisconsin Biotechnology Association and a co-author condemning Wisconsin's U.S. Senator Feinstein's opposition to this form of biotechnology (Timmins & George, 1993), also largely on grounds of encouraging economic competitiveness.

Grocery chains and dairy cooperatives, caught in the middle of this complex debate, sometimes opted for caution. But in the midwest, the biotech industry was widely reported to be pressuring co-ops refusing to accept milk produced using the bioengineered hormone. Meanwhile in Wisconsin, at least one local school board banned the milk from treated cows from its school lunches (as reported by the *St. Louis Post-Dispatch*, hometown paper to those at Monsanto headquarters; see Steyer, 1994). Wisconsin was generally cited as the hotbed of opposition. For example, the *Milwaukee Journal Sentinel* reported a 1995 survey of Wisconsin dairy farmers showing only 6.6% were using the product (Bergquist, 1996).

Whatever the eventual impact on the Wisconsin dairy industry, at the national and even local levels resistance eventually faded from view until two Florida journalists were fired in 1997 for allegedly refusing to frame their investigative story in terms acceptable to industry. They won their suit against Fox in August 2000, but with little or no noticeable effect on the national biotechnology dialogue (Cotts, 2000). Although citizen involvement in evaluation of biotechnology research in Cambridge was portrayed as a model of rational citizen involvement in policymaking, Midwestern farm resistance attracted little praise from either side of the controversy. When it did appear in the national press, it was typically portrayed as a marketing issue, not a farming issue. Limited recognition that farmers had a role in this kind of decision making may have played some part in this (Michael, 1996).

Local objections of a different sort arose a few years later over a $50 million research alliance between the University of California at Berkeley and the Swiss biotechnology company Novartis. Although early reports seemed quite positive, stressing the benefits to the university, objections quickly arose in publications ranging from the student *Daily Californian* to the *San Francisco Chronicle*. This controversy, which revolved around Novartis' guaranteed first crack at marketing any commercially viable results produced by the university's Department of Plant and Microbial Biology, was at least mentioned in the *Los Angeles Times* and received some critical national media attention (e.g., in a *U.S. News and World Report* piece; see Petit, 1998). The *Chronicle* later reported that "protests and pie throwing" accompanied the ceremonial signing (Burress, 1998; also reported in *The Washington Times* by Elias, 1998). The agreement went through despite concerns both internal and external to the institution, and media attention largely died away.

Although the growth hormone controversy upset the dairy industry, this research deal also got a lot of attention in the higher education press, with pieces on the controversial aspects appearing in the *Chronicle of Higher Education* (Blumenstyk, 1998) and the National Center for Public Policy and Higher Education's *National Crosstalk* (Irving, 1999; Rosenzweig, 1999). Ironically, criticism printed in *Science* magazine focused primarily on concerns brought up in a report co-authored by a Pfizer CEO who preferred a less publicized Monsanto–Washington University arrangement to that of Novartis and Berkeley ("Partnership Perils," 2001). The interesting parallel to the somatotropin case is that in both cases economic policy and its impact, in addition to health issues and other matters susceptible to scientific evidence, were central to the debates. The additional arguments may not have been widely understood, however, outside the respective industries, whether within small-farm dairies or research universities.

The range of possible objections and the way local media framing of them can evolve is nicely illustrated by a series in the Claremont, California, *Courier* by reporter and city editor Gary Scott.[11] The original story published on April 10, 1999, concerned local community objections to plans by the Keck Graduate Institute of Applied Life Sciences to locate its building on a piece of land that was over 11 acres at the Bernard Biological Field Station. The conflict was primarily over local land use, although there was some discussion regarding the relative merits of field and laboratory as research and teaching sites. Over time, concerns represented in the media accounts evolved. What began as consideration of the specific land use planning issue soon encompassed possible violations of local Native Americans' hopes for—if not legal rights to—the area, corporate domination of the biotechnology research agenda, allegations of abuse of power, and the implications of the human genome project for our understanding of the nature and meaning of human life.

Thus, what began as a very local struggle over an environmental impact assessment gradually became indistinguishable from broader national and international debates over economics and ethics. It is difficult not to conclude that the local media attention these broader issues garnered would never have taken place without the concern over local planning issues, making this a case in which the definition of the news agenda seemed to simultaneously move from the bottom up and from the top down. The protests in Claremont eventually resulted in 15 student arrests and at least two articles in the *Los Angeles Times* the following spring. Yet as national discussion of biotechnology and its impact remained inchoate, the incident was unlikely to be seen as linked to broadly shared policy concerns. Instead this was seen as only a local issue and not something that should direct national policy.

In 1999, Percy Schmeiser, a farmer in central Canada, was sued by Monsanto for growing unauthorized bioengineered canola plants on his farm—plants he claimed sprouted from seed fertilized by pollen drifting on the wind from neighborhood farms. The image of the lone independent farmer stubbornly holding out against attack by the giant agribusiness company was striking enough to capture the attention of award-winning *Washington Post* science writer Rick Weiss (1999), although Schmeizer eventually lost the suit (appeals continue). This is the exception that proves the rule, however. These examples are only a few cases among many, many scattered incidents of protest, ranging from supermarket picketing and soybean dumping in Maine to the late-night bombing of a research lab at Michigan State University to what appears to be a growing number of incidents of crop destruction in various locations (Stape, 2001).

Generally speaking, with the exception of the furor over Dolly the sheep, local protest in the United States has done little to upset the conventional wisdom that not many people—at least not many sane or "normal" people—have questions here about genetic engineering, whether medical or agricultural. Opposition and protest to genetically engineered foods exist in the United States just as they do in Europe, but nationally visible reflections of these dynamics have been rare until recently. Rather, such protests are widely perceived as the actions of a small group of radical, unbalanced, chronic malcontents. Although this may be justifiable in some individual cases, such an interpretation belies the widespread and varied nature of the objections being raised and the conservative character of many of those raising them. This includes everyone from experts on higher education policy to smaller scale dairy farmers in the upper midwest to ordinary residents of Claremont, California—and even some members of the academic research community.

The Spiral of Silence and Risk Perception

Perhaps the most disturbing part of the biotechnology debate in the United States has been its demonstration of the influence of large-institution dominance over the way in which news accounts of technological controversy unfold, although it is reassuring to some to note that the biotechnology industry has failed to maintain this control entirely. However, without the appearance of dissent from outside its borders and breakthrough events at the local level, it is possible that the biotech spiral of silence would never have been challenged in any important way. As both news media and industrial science become more globalized, what are the implications for the future?

Functionalist analyses of the political role of the U.S. media usually focus on its "watchdog" role and the "marketplace of ideas" concept through which democratic processes are supposed to distill truth, or at least wisdom. In this approach, the media are expected to function as moderators of information that treat newsmakers and news consumers in the same light. As explored by Best and Kellner (chap. 8, this volume), these roles are always more difficult to achieve for highly technical issues in which there is a recognized—and legitimate—role for scientific expertise to play. In cases of such controversy, the scientific voice can carry enormous rhetorical power even when articulating arguments are emotional rather than technical (Waddell, 1990); clear lines between arguments having to do with ethics, values, and democratic choice are difficult or impossible to separate cleanly from arguments having to do with scientific facticity. As a result, dissent may be easier to suppress in cases of technological controversy than in cases of controversy of other types. As the number of mainstream independent news voices in the world shrinks, the prospects for global hegemony—not just national hegemony—increase. The internet has made this a more tenuous position, although this assumes consumer access, and there are those who feel that the internet is, or will be, controlled by the same organizations that control the large communication businesses (e.g., Neuman, 1991).

It appears as if public concern over biotech in the United States has begun to influence federal policy in important ways. At the time of this writing, for example, it seems likely that some form of food labeling for GM content will emerge as U.S. policy in response to widely demonstrated consumer reservations, although the Bush administration is likely to delay or diminish this change. However, when the news media reported general public complacence about biotechnology, the policy community seemed ready and willing to accept it. Under what circumstances does the spiral of silence prevail, and under what circumstances is it broken? For issues like biotechnology, people who are not scientists may have little reason to question scientific authority or articulate misgivings unless they have some means of exposure to alternative interpretations and nonmainstream positions. Readers and audiences often question science on the basis of life experience regardless of whether media stories introduce critiques (Priest, 1995). Local controversies that may never reach the national news agenda have likely been an important source of oppositional readings in the area of biotechnology.

CONCLUSIONS

Resistance to biotechnology and genetic engineering is not relegated to Green-controlled European nations or "uneducated" developing nations. There has been a great deal of resistance to these new technologies within the United States; the purpose of this chapter has been to offer a possible explanation for that resistance. We began by noting that the coverage offered by the national press, represented by *The New York Times* and *The Washington Post*, did not accurately reflect the opinion held by the public until recently. A suite of factors appears to have been important to this change.

First, information subsidies offered by those with a vested interest in the biotechnology sector outweighed oppositional voices until the point at which other factors, such as a series of dramatic breakthrough events, began to shift the costs of ignoring the latter for reporters and editors at the national level. Although the spiral of silence was clearly under attack as early as the 1970s, it is only recently that oppositional voices have begun to receive space in the agenda-setting media of the national prestige press.

Second, the breakthrough events that have successfully influenced the nature of media reportage in the United States dealing with public opinion on biotechnology-related issues have taken place at both the supra- and subnational levels. European concerns finally entered U.S. public discourse when it began to influence opportunities for export of American foods and when health concerns became very widespread. Local opposition to biotechnology, generated in response to locally perceived threats in the United States, began to affect the national media agenda as the events and their effects on public opinion and decision making began to accrue over time. Local opposition is most successful in affecting the national news media when it captures the imagination in unusual ways, whether by introducing new decision-making processes or using high drama in expressions of dissent.

Third, shifts in the framing of biotechnology by both public and media seem to have made a difference. Although the biotechnology industry successfully encouraged media framing of biotechnology as a premiere example of "progress" for many years, local opposition to biotechnology often stemmed from and in turn created media frames that focused instead on biotechnology as "risk." Risk perception may or may not be related to scientific evidence; global warming is a threat widely accepted by scientists, yet conservative politics insists on a reading that rejects it. Evolutionary theory explains important biological and geological evidence, yet fundamentalist Christianity claims epistemological equivalency. Some African leaders have been portrayed as rejecting the HIV theory of AIDS—a perspective that does not bode well for attempts to control the epidemic there. It would be naive to propose that the popular wisdom about issues related to science and technology is always best. Yet ceding control of science-related policy decisions exclusively to the technocratic elite does not bode well for democracy either.

In the case of biotechnology, industrial interests have failed to suppress dissent completely. Although much U.S. policy, taken as a whole, continues to support commercial interests, popular opinion is having an impact. For scientific controversies, the creation of dissent may be an easier goal to accomplish than the maintenance of ideological domination—a dynamic for which we should sometimes be grateful. Lay wisdom in the United Kingdom rejected assurances that mad cow was harmless to humans; lay wisdom in the United States rejected assurances that the nuclear near-catastrophe at Three Mile Island could never happen; lay wisdom on both sides of the Atlantic may lean increasingly in the direction of precaution, albeit in different ways. Yet experts' fears about public rejection of science and technology are not devoid of reason either and could make the achievement of better environmental protection, more sustainable agriculture, and improved public health more difficult.

We can only hope that lay wisdom, in an open information environment, maintains its resilience to institutional manipulation of whatever kind, given the special challenges posed by technically complex controversies. Science education and science communication have contributions to make toward this end that are much more challenging than simply getting the facts straight or even (in the case of journalism) resisting domination by sources with vested interests in positive public attitudes toward various forms of science. The goal should not be the creation of consent, but it is not necessarily the creation of dissent either. Rather, the goal is the creation of the appropriate conditions for productive public dialogue that acknowledges, but does not assume, the expert point of view and the role of expert testimony. This includes broad public recognition of both the power and the ultimate uncertainty of science.

Mechanisms for citizen input into policy formation in the United States are limited. The prospect is fairly daunting. The national news agenda and the ways issues are defined within it are important factors in public policy determination, if not strong determinants of public opinion formation. Both the political left and the political right in this country are associated in part with rejection of science and technology. From the religious right, anti-abortionists oppose stem cell research because they associate the availability of embryos with pregnancies that have been artificially interrupted. At the other end of the spectrum, environmental advocates may reject forms of science and technology they see as unsustainable. As the biotechnology debate demonstrates, repressed resistance reemerges, although its expression at the national level and in national policy can be suppressed or delayed.

ENDNOTES

1. An early version of this chapter was presented at the May 2001 meeting of the International Communication Association in Washington, DC.

2. By *biotechnology* we mean recombinant DNA and closely related technologies as used in both medical and agricultural applications. Although commentators often take care to distinguish between the two, and they do sometimes engender somewhat different public responses, both media accounts and public opinion tend to blur them together, and the underlying science is often shared.

3. This argument is not new, but was raised decades ago with respect to the emergence of violent protests over the Vietnam War, which may have occurred in part because news media steadfastly ignored the existence of dissent earlier.

4. Separately, we are participating in a large international study designed to examine the similarities and differences between European and U.S. press coverage more fully (Gaskell et al., 2001).

5. The authors would like to thank Edna Einsiedel, who is conducting similar research in Canada, for sharing some of her work with the researchers in personal communication contexts.

6. The authors would like to thank Harry Perlstadt for pointing out these other controversies.

7. Coding categories used for our content analysis as reported here and elsewhere are based on a scheme developed by the international group cited previously.

8. It should be noted that, although *frame* refers to the main slant of the article in terms of focus, it does not necessarily correlate with the overall tone of the article. In other words, an article may discuss new genetic diagnostics, but focus on concerns with insurance coverage. This would be a progressive frame, but a negatively toned article.

9. It is important not to forget that the reasons for the emergence of this story are problematic. GMOs, including domestic agricultural animals such as cattle, were under development for years before this particular breakthrough became news. The technology to clone

human beings in a different sense—through embryo splitting rather than somatic cell repro-gramming—has also been available for some time.

10. A bioengineered rennet substitute was approved in 1990; Calgene requested FDA review of its Flavr Savr tomato in 1991, but this was not completed until 1994. Neither product seems to have been the target of particularly widespread objections, whether local or national, in comparison with the growth hormone product.

11. Material and supporting *Courier* articles provided by Mr. Scott via personal communication, July 2000.

8

Biotechnology, Democracy, and the Politics of Cloning[1]

Steven Best
University of Texas–El Paso

Douglas Kellner
UCLA

O, wonder!
How many goodly creatures are there here!
How beauteous mankind is!
O brave new world
That has such people in't.

—William Shakespeare, *The Tempest*

We're ready to go because we think that the genie's out of her bottle.
—Dr. Panos Zavos

Anyone who thinks that things will move slowly is being very naive.
—Lee Silver, Molecular Biologist

As we move into a new millennium fraught with terror and danger, a global postmodern cosmopolis is unfolding in the midst of rapid evolutionary and social changes co-constructed by science and technology. We are quickly morphing into a new biological and social existence that is ever more mediated and shaped by computers, mass media, and biotechnology, all driven by the logic of capital and a powerful, emergent technoscience. In this global context, science is no longer merely an interpretation of the natural and social worlds. Rather it has become an active force in changing them and the very nature of life. In an era where life can be created and resigned in a petri dish and genetic codes can be edited like a digital text, the distinction between *natural* and *artificial* has become confused and confounded.

The new techniques of manipulation call into question existing definitions of life and death, demand a rethinking of fundamental notions of ethics and moral value, and pose unique challenges for democracy.

As technoscience develops by leaps and bounds, and as genetics rapidly advances, the science–industry complex has come to a point where it is creating new transgenic species and is rushing toward a posthuman culture that unfolds in the ever more intimate merging of technology and biology. The posthuman involves both new conceptions of the human in an age of information and communication, and new modes of existence as flesh merges with steel, circuitry, and genes from other species. Exploiting more animals than ever before, technoscience intensifies research and experimentation into human cloning. This process is accelerated because genetic engineering and cloning are developed for commercial purposes, anticipating enormous profits on the horizon for the biotech industry. Consequently, all natural reality—from microorganisms and plants to animals and human beings—is subject to genetic reconstruction in a commodified "Second Genesis."

At present, the issues of cloning and biotechnology are being heatedly debated in the halls of science, in political circles, among religious communities, throughout academia, and more broadly in the media and public spheres. Not surprisingly, the discourses on biotechnology are polarized. Defenders of biotechnology extol its potential to increase food production and quality, cure diseases and prolong human life, and better understand human beings and nature to advance the goals of science. Its critics claim that genetic engineering of food will produce Frankenfoods that pollute the food supply with potentially harmful products; that biotechnology out of control could devastate the environment, biodiversity, and human life; that animal and human cloning will breed monstrosities; that a dangerous new eugenics is on the horizon; and that the manipulation of embryonic stem cells violates the principle of respect for life and destroys a bona fide human being.

Interestingly, the same dichotomies that have polarized information-technology discourses into one-sided technophobic and technophilic positions are reproduced in debates over biotechnology. Just as we have argued that critical theories of technology are needed to produce more dialectical perspectives that distinguish between positive and negative aspects and effects of information technology, so too would we claim that similar multiperspectival approaches are required to articulate the potentially beneficial and perhaps destructive aspects of biotechnology. Indeed current debates over cloning and stem cell research suggest powerful contradictions and ambiguities in these phenomena that render one-sided positions superficial and dangerous. Parallels and similar complexities in communication and biotechnology are not surprising given that information technology provides the infrastructure to biotechnology that has been constituted by computer-mediated technologies involved in the Human Genome Project. Conversely, genetic science is being used to push the power and speed of computers through phenomena such as biological chips (see Best & Kellner, 2001).

As the debates over cloning and stem cell research indicate, issues raised by biotechnology combine research into the genetic sciences, perspectives and contexts articulated by the social sciences, and the ethical and anthropological concerns of philosophy. Consequently, we argue that intervening in the debates over biotechnology require supradisciplinary critical philosophy and social theory to illuminate the problems and their stakes. In addition, debates over cloning and stem cell research raise exceptionally important challenges to a democratic politics of communication. Biotechnology is thus a critical flashpoint for democratic theory and practice. It underscores the need for more widespread knowledge of important scientific issues, participatory debate, consensus, and regulation concerning new developments in the biosciences, which have such high economic, political, and social consequences.

More specifically, we demonstrate problems with the cloning of animals that for now render the cloning of humans unacceptable. In our view, human cloning constitutes a momentous route to the posthuman—a leap into a new stage of history with significant and potentially disturbing consequences. We also take on arguments for and against stem cell research and contend that it contains positive potential for medical advances that should not be blocked by problematic conservative positions. Nonetheless, we believe that the entire realm of biotechnology is fraught with dangers and problems that require careful study and democratic debate. The emerging genomic sciences should thus be undertaken by scientists with a keen sense of responsibility and accountability and be subject to intense public scrutiny and open discussion. Finally, in the light of the dangers and potentially deadly consequences of biotechnology, we maintain that embracing its positive potential can be realized only in a new context of genuine social democracy and new sensibilities toward nature.

BRAVE NEW BARNYARD: THE ADVENT
OF ANIMAL CLONING

The idea is to arrive at the ideal animal and repeatedly copy it exactly as it is.
—Dr. Mark Hardy

We are up to our ears in [animal] clones.
—Michael Bishop, President of Infigen Inc.

From its entrenched standpoint of unqualified human superiority, science typically first targets objects of nature and animals with its analytic gaze and instruments. The current momentous turn toward cloning is largely undertaken by way of animals, yet some scientists have already directly focused on cloning human beings. Although genetic engineering creates new transgenic species by inserting the gene from one species into another, cloning replicates cells to produce identical copies of a host organism by inserting its DNA into an enucleated egg. In a potent combination, genetic engineering and cloning technologies are used together, first, to custom design a transgenic animal to suit the needs of science and industry (the distinction is irrevocably blurred), and, second, to mass reproduce the hybrid creation endlessly for profitable peddling in medical and agricultural markets.

Cloning is a return to asexual reproduction and bypasses the caprice of the genetic lottery and random shuffling of genes. It dispenses with the need to inject a gene into thousands of newly fertilized eggs to get a successful result. Rather, much as the printing press replaced the scribe, cloning allows mass reproduction of a devised type, and thus opens genetic engineering to vast commercial possibilities. Life-science companies are poised to make billions of dollars in profits as numerous organizations, universities, and corporations move toward cloning animals and human stem cells, and patenting the methods and results of their research.

To date science has engineered thousands of varieties of transgenic animals and has cloned sheep, calves, goats, bulls, pigs, and mice. Although still far from precise, cloning nevertheless has become routine. What is radically new and startling is not cloning itself; since 1952 scientists have replicated organisms from embryonic cells. Rather the new techniques of cloning, or nuclear somatic transfer, from adult mammal body cells constitutes a new form of human reproduction. These methods accomplish what scientists long considered impossible—reverting adult (specialized) cells to their original (non-specialized) embryonic state where they can be reprogrammed to form a new organism. In effect this startling process creates the identical twin of the adult that provided the original donor cell. This technique was used first to create Dolly and subsequently all of her varied offspring.

Dolly and Her Progeny

Traditionally, scientists considered cloning beyond the reach of human ingenuity. But when Ian Wilmut and his associates from the Roslin Institute near Edinburgh, Scotland, announced their earth-shattering discovery in March 1997, the impossible appeared in the form of a sheep named Dolly, and a natural law had been broken. Dolly's donor cells came from a 6-year-old Finn Dorset ewe. Wilmut starved mammary cells in a low-nutrient tissue culture where they became quiescent and subject to reprogramming. He then removed the nucleus containing genetic material from an unfertilized egg cell of a second sheep, a Scottish blackface, and in a nice Frankenstein touch fused the two cells with a spark of electricity. After 277 failed attempts, the resulting embryo was then implanted into a third sheep—a surrogate mother who gave birth to Dolly in July 1996.[2]

Many critics said Dolly was either not a real clone or was just a fluke. Yet less than 2 years after Dolly's appearance, scientists had cloned numerous species, including mice, pigs, cows, and goats, and had even made clones of clones of clones, producing genetic simulacra in mass batches as Aldous Huxley (1989a [1932]) envisioned happening to human beings in *Brave New World.* The commercial possibilities of cloning animals were dramatic and obvious for all to behold. The race was on to patent novel cloning technologies and the transgenic offspring they would engender.

Animals are being designed and bred as living drug and organ factories as their bodies are disrupted, refashioned, and mutilated to benefit meat and dairy industries. Genetic engineering is employed in biomedical research by infecting animals with diseases that become part of their genetic makeup and are transmitted to their offspring, as in the case of researchers trying to replicate the effects of cystic fibrosis in sheep. Most infamously, Harvard University, with funding from Du Pont, has patented a mouse— OncoMouse—that has human cancer genes built into its genetic makeup and are expressed in its offspring (Haraway, 1997).

In the booming industry of *pharming* (pharmaceutical farming), animals are genetically modified to secrete therapeutic proteins and medicines in their milk. The first major breakthrough came in January 1998 when Genzyme Transgenics created transgenic cattle named George and Charlie. The result of splicing human genes and bovine cells, they were cloned to make milk that contains human proteins such as the blood-clotting factor needed by hemophiliacs. Co-creator James Robl said, "I look at this as being a major step toward the commercialization of this [cloning] technology."[3]

In early January 2002, the biotech company PPL announced that they had just cloned a litter of pigs that could aid in human organ transplants— on the eve of the publication of an article by another company Immerge Bio Therapeutics, which claimed it had achieved a similar breakthrough.[4] The new process involved creation of the first knockout pigs, in which a single gene in pig DNA is knocked out to eliminate a protein present in pigs that is usually violently rejected by the human immune system. This meant that a big step could be made in the merging of humans and animals and creating animals as harvest machines for human organs.

Strolling through the Brave New Barnyard, one can find incredible beings that appear normal, but are genetic satyrs and chimera. Cows generate lactoferrin, a human protein useful for treating infections. Goats manufacture antithrombin III, a human protein that can prevent blood clotting, and serum albumin, which regulates the transfer of fluids in the body. Sheep produce alpha antitrypsin, a drug used to treat cystic fibrosis. Pigs secrete phytase, a bacterial protein that enables them to emit less of the pollutant phosphorous in their manure; and chickens make lysozyme, an antibiotic, in their eggs to keep their own infections down.

BioSteel presents an example of the bizarre wonders of genetic technology that points to the erasure of boundaries between organic and inorganic matter, as well as between different species. In producing this substance, scientists have implanted a spider gene into goats so that their milk produces a super-strong material—BioSteel—that can be used for bulletproof vests, medical supplies, and aerospace and engineering projects. To produce vast quantities of BioSteel, Nexia Biotechnologies intends to house thousands of goats in 15 weapons-storage buildings, confining them in small holding pens.[5]

Animals are genetically engineered and cloned for yet another reason—to produce a stock of organs for human transplants. Given the severe shortage of human organs, thousands of patients every year languish and die before they can receive a healthy kidney, liver, or heart. Rather than encouraging preventive medicine and finding ways to encourage more organ donations, medical science has turned to xenotransplantation and has begun breeding herds of animals (with pigs as a favored medium) to be used as organ sources for human transplantation.

Clearly, this is a hazardous enterprise due to the possibility of animal viruses causing new plagues and diseases in the human population (a danger that also exists in pharmaceutical milk). For many scientists, however, the main concern is that the human body rejects animal organs as foreign and destroys them within minutes. Researchers seek to overcome this problem by genetically modifying the donor organ so that they knock out markers in pig cells and add genes that make their protein surfaces identical to those in humans. Geneticists envision cloning entire herds of altered pigs and other transgenic animals so that an inexhaustible warehouse of organs and tissues would be available for human use. In the process of conducting experiments such as transplanting pig hearts modified with a human gene into the bodies of monkeys, companies such as Imutran have caused horrific suffering, with no evident value to be gained given the crucial differences among species and introducing the danger of new diseases into human populations.[6]

As if billions of animals were not already exploited enough in laboratories, factory farms, and slaughterhouses, genetic engineering and cloning exacerbate the killing and pain with new institutions of confinement and bodily invasion that demand millions and millions more captive bodies. Whereas genetic and cloning technologies in the cases described at least have the potential to benefit human beings, they have also been appropriated by the meat and dairy industries for purposes of increased profit through the exploitation of animals and biotechnology. It is the nightmarish materialization of the H. G. Wells scenario where, in his prophetic 1904 novel *The Food of the Gods*, scientists invent a substance that prompts every living being that consumes it to grow to gargantuan proportions.[7] Having located the genes responsible for regulating growth and metabolism, university and corporate researchers immediately exploited this knowledge for profit. Thus, for the glories of carnivorous consumption, corporations such as MetaMorphix and Cape Aquaculture Technologies have created giant pigs, sheep, cattle, lobsters, and fish that grow faster and larger than the limits set by evolution.

Amid the surreality of Wellsian gigantism, cattle and dairy industries are engineering and cloning designer animals that are larger, leaner, and faster growing value producers. With synthetic chemicals and DNA alteration, farmers can produce pigs that mature twice as fast and provide at least twice the normal amount of sows per litter as they eat 25% less feed and cows that produce at least 40% more milk. Since 1997, at least one country—Japan—has sold cloned beef to its citizens.[8] Yet there is strong reason to believe that U.S. consumers—already a nation of guinea pigs in their consumption of genetically modified foods—have eaten cloned meat and dairy products. For years, corporations have cloned farmed animals with the express purpose of someday introducing them to the market, and insiders claim many already have been consumed.[9] The U.S. National Institute of Science and Technology has provided two companies—Origen Therapeutics of California and Embrex of North Carolina—with almost $5 million to fund research into factory farming billions of cloned chickens for consumption.[10] With the U.S. Food and Drug Administration (FDA) pondering whether to regulate cloned meat and dairy products, it is a good bet that they are many steps behind an industry determined to increase its profits through biotechnology. The future to come seems to be one of cloned humans eating cloned animals.

Although anomalies such as self-shearing sheep and broiler chickens with fewer feathers have already been assembled, some macabre visionaries foresee engineering pigs and chickens with flesh that is tender or can be easily microwaved, and chickens that are wingless so they will not need bigger cages. The next step would be to just create and replicate animals' torsos—sheer organ sacks—and dispense with superfluous heads and limbs. In fact scientists have already created headless embryos of mice and frogs in grotesque manifestations of the kinds of life they can now construct at will.

Clearly, there is nothing genetic engineers will not do to alter or clone an animal. Transgenic artist Eduardo Kac, for instance, commissioned scientists at the National Institute of Agronomic Research in France to create Alba, a rabbit that carries a fluorescent protein from a jellyfish and thus glows in the dark. This experiment enabled Kac to demonstrate his supremely erudite postmodern thesis that, "genetic engineering [is] in a social context in which the relationship between the private and public spheres are negotiated"![11] Although millions of healthy animals are euthanized every year in U.S. animal "shelters," corporations are working to clone animals either to bring them back from the dead or prevent them from *dying* (such as in the Missyplicity Project, initiated by the wealthy owners of a dog who want to keep her alive indefinitely).[12] Despite alternatives to coping with allergy problems and the dangers with cloning animals, Transgenic Pets LLC is working to create transgenic cats that are allergen-free.[13] It is time to examine concretely what cloning means for animal existence.

Transgenic Travesties

The agricultural use of genetics and cloning has produced horrible monstrosities. Transgenic animals are often born deformed and suffer from fatal bleeding disorders, arthritis, tumors, stomach ailments, kidney disease, diabetes, inability to nurse and reproduce, behavioral and metabolic disturbances, high mortality rates, and Large Offspring Syndrome. To genetically engineer animals for maximal weight and profit, a Maryland team of scientists created the infamous "Beltway pig" afflicted with arthritis, deformities, and respiratory disease. Cows engineered with bovine growth hormone (rBGH) have mastitis, hoof and leg maladies, reproductive problems, and numerous abnormalities, and they die prematurely. Giant supermice endure tumors, damage to internal organs, and shorter life spans. Numerous animals born from cloning are missing internal organs such as hearts and kidneys. A Maine lab specialized in breeding sick and abnormal mice who go by names such as Fathead, Fidget, Hairless, Dumpy, and Greasy. Similarly, experiments in the genetic engineering of salmon have led to rapid growth and various aberrations and deformities, with some growing up to 10 times their normal body weight (Fox, 1999). Cloned cows are 10 times more likely to be unhealthy as their natural counterparts. After 3 years of efforts to clone monkeys, Dr. Tanja Dominko fled from her well-funded Oregon laboratory. Telling cautionary tales of the "gallery of horrors" she experienced, Dominko said that 300 attempts at cloning monkeys produced nothing but freakishly abnormal embryos that contained cells either without chromosomes or with up to nine nuclei.[14]

For Dominko, a "successful" clone like Dolly is the exception, not the rule. But even Dolly is inexplicably overweight, and there was evidence in May 1999 that she may be susceptible to premature aging. On January 4, there were reports that Dolly has arthritis, and her creator Ian Wilmut said on a BBC broadcast: "There is no way of knowing if this is due to cloning or whether it is a coincidence." Moreover, cloned mice have also become extremely obese, and cloned cows have been born with abnormally large hearts and lungs.

A report from newscientists.com argues that genes are disrupted when cultured in a lab, and this explains why so many cloned animals die or are grossly abnormal. On this account, it is not the cloning or IVF process that is at cause, but the culturing of the stem cells in the lab, creating major difficulties in cloning since so far there is no way around cloning through cultured cells in laboratory conditions.[15]

A team of U.S. scientists at the MIT Whitehead Institute examined 38 cloned mice and learned that even clones that look healthy suffer genetic maladies, and scientists found that mice cloned from embryonic stem cells had abnormalities in the placenta, kidneys, heart, and liver. They feared that the defective gene functioning in clones could wreak havoc with organs and trigger foul-ups in the brain later in life and that embryonic stem cells are highly unstable.[16] "There are almost no normal clones," study author and MIT biology professor Rudolf Jaenisch explained. Jaenisch claimed that only 1% to 5% of all cloned animals survive, and even those that survive to birth often have severe abnormalities and die prematurely.[17]

As we argue later, these risks make human cloning a deeply problematic undertaking. Pro-cloning researchers claim that the "glitches" in animal cloning can be worked out eventually. In January 2001, for example, researchers at Texas A&M University and the Roslin Institute claimed to have discovered a gene that causes abnormally large cloned fetuses—a discovery they believe will allow them to predict and prevent this type of mutation. It is conceivable that someday science will work out the kinks, but for many critics this assumes that science can master what arguably are inherent uncertainties and unpredictable variables in the expression of genes in a developing organism. A recent study showed that some mouse clones seem to develop normally until an age the equivalent of 30 years for a human being; then there is a spurt of growth and they suddenly become obese.[18] Mark Westhusin, a cloning expert at Texas A&M, pointed out that the problem is not that of genetic mutation, but of genetic expression—that genes are inherently unstable and unpredictable in their functioning. Another report indicates that a few misplaced carbon atoms can lead to cloning failures.[19] Thus, any small errors in the cloning process could lead to huge disasters, and the prevention of all such small errors seems to presume something close to omniscience.[20]

Yet the matter has become controversial because other scientists are now claiming that they have produced "normal" cloned animals. In June 2001, the University of Georgia announced that it had successfully cloned eight cattle using a new and improved method that allegedly raises the survival rate from 5% to 14.3%. Still this means that only one out of seven of the cloned cattle will live using current technology. On November 23, 2001, *Sci-*

ence published a study by the Mayo Clinic, the University of Pennsylvania, and three companies involved in animal cloning, including Advanced Cell Technology (ACT), which claimed that 24 cow clones were reported to be "normal in every way" after several years of experimentation. The company had created 500 cloned embryos implanted into 250 cows. Of those only 110 became pregnant, 80 miscarried, 30 survived to birth, and 24 survived to adulthood. This is not exactly a success story, but it does not preclude the possibility that science might be improving in its ability to clone. Critics like Jaenisch, however, question the claim to normality and argue that the "tests are very superficial" and genetic problems could turn up later (*The Washington Post*, Nov. 23, 2001). Moreover, several of the scientists who authored the study have financial stakes in the animal cloning industry and so have a vested interest in disseminating junk science and good PR releases—a tactic not beneath the "objectivity" of corporations such as ACT that willfully implode the boundary between science and publicity. Indeed in the highly competitive cloning marketplace, where companies are scrambling to patent the first major breakthrough in stem cell research, PR and the manipulation of media are lab tools as basic as a microscope.

Although most scientists are opposed to cloning human beings (rather than stem cells) and decry it as unacceptable, none condemns the suffering caused to animals or positions animal cloning research as morally problematic, and animal rights groups so far have been excluded from the debate. Quite callously and arbitrarily, for example, Jaenisch proclaimed, "You can dispose of these animals, but tell me—what do you do with abnormal humans?"[21] The attitude that animals are disposable is a good indication of the problems inherent in the mechanistic science that still prevails and a symptom of callousness toward human life that is worrisome.

Despite the claims of its champions, the genetic engineering of animals is a radical departure from natural evolution and traditional forms of animal breeding, whereas human cloning takes the postmodern adventure of rapid technoscientific change into a new and, to many, frightening posthuman realm that begins to redesign the human body and genome (see Best & Kellner, 2001). Cloning involves manipulation of genes rather than whole organisms. Moreover, scientists engineer change at unprecedented rates and can create novel beings across species boundaries that were previously unbridgeable. Ours is a world in which cloned calves and sheep carry human genes, human embryo cells are merged with enucleated cows' eggs, monkeys and rabbits are bred with jellyfish DNA, a surrogate horse gives birth to a zebra, an ordinary dairy cow spawns an endangered gaur, and tiger cubs emerge from the womb of an ordinary housecat.

The ability to clone a desired genetic type brings the animal kingdom into entirely new avenues of exploitation and commercialization. From the new scientific perspective, animals are framed as genetic information that can be edited, transposed, and copied endlessly. Pharming and xenotransplantation build on the system of factory farming that dates from the postwar period and are based on the confinement and intensive management of animals within enclosed buildings that are prison houses of suffering.

The proclivity of the science–industry complex to instrumentalize animals as nothing but resources for human use and profit intensifies in an era in which genetic engineering and cloning are perceived as sources of immense profit and power. Still confined for maximal control, animals are no longer seen as whole species, but rather as carriers of genetic information to be manipulated for any purpose.

Weighty ethical and ecological concerns in the new modes of animal appropriation are largely ignored; animals are still framed in the 17th-century Cartesian worldview that views them as nonsentient machines. As Rifkin (1999) puts it,

> Reducing the animal kingdom to customized, mass-produced replications of specific genotypes is the final articulation of the mechanistic, industrial frame of mind. A world where all life is transformed into engineering standards and made to conform to market values is a dystopian nightmare, and needs to be opposed by every caring and compassionate human being who believes in the intrinsic value of life. (p. 35)[22]

Patenting of genetically modified animals has become a huge industry for multinational corporations and chemical companies. PPL Therapeutics, Genzyme Transgenics, Advanced Cell Technology, and other enterprises are issuing broad patent claims on methods of cloning nonhuman animals. PPL Therapeutics, the company that invented Dolly, has applied for the patents and agricultural rights to the production of all genetically altered mammals that could secrete therapeutic proteins in their milk. Nexia Biotechnologies obtained exclusive rights to all results from spider silk research. Patent number 4,736,866 was granted to Du Pont for Oncomouse, which the Patent Office described as a new "composition of matter." Infigen holds a U.S. patent for activating human egg division through any means (mechanical, chemical, or otherwise) in the cloning process.

Certainly, genetics does not augur solely negative developments for animals. Given the reality of dramatic species extinction and loss of biodiversity, scientists are collecting the sperm and eggs of endangered species like the giant panda to preserve them in a frozen zoo. It is indeed exciting to ponder the possibilities of a Jurassic Park scenario of reconstructed extinct species (e.g., scientists recently uncovered the well-preserved remains of a Tasmanian tiger and a woolly mammoth). In 2001, European scientists cloned a seemingly healthy mouflon lamb, a member of an endangered species of sheep, and ACT produced the first successful interspecies clone when a dairy cow gave birth to a gaur, an endangered wild ox native to Southeast Asia (although it died of an infection only 2 days later). Currently, working with preserved tissue samples, ACT is working to bring back from extinction the last bucardo mountain goat, which was killed by a falling tree in January 2000.[23]

Critics, however, dismiss this as a misguided search for a technofix that distracts focus from the real problem of preserving habitat and biodiversity. Even if animals could be cloned, there is no way to clone habitats lost forever to chainsaws and bulldozers. Moreover, the behaviors of cloned animals would unavoidably be altered, and they would end up in zoos or exploitative entertainment settings where they exist as spectacle and simulacra. Animals raised through interspecies cloning such as the gaur produced by ACT will not have the same disposition as if raised by their own species and so for other reasons will not be less than real. Additionally, there is the likelihood that genetic engineering and cloning would aggravate biodiversity loss to the extent that it creates monolithic superbreeds that could crowd out other species or be easily wiped out by disease. There is also great potential for ecological disaster when new beings enter an environment, and genetically modified organisms are especially unpredictable in their behavior and effects.

Still, cloning may prove a valuable tool in preserving what can be salvaged from the current extinction crisis. Moreover, advances in genetics may also bypass and obviate pharming and xenotransplantation through the use of stem cell technologies that clone human cells, tissues, or perhaps even entire organs and limbs from human embryos or an individual's own cells. Successful stem cell technologies could eliminate at once the problem of immune rejection and the need for animals. There is also the intriguing possibility of developing medicines and vaccines in plants, rather than animals, thus producing a safer source of pharmaceuticals and neutraceuticals and sparing animals suffering. However, none of these promises brightens the dark cloud cloning casts over the animal kingdom or dispels the dangers of the dramatic alteration of human life.

CLONES R' US: THE PORTENT OF HUMAN REPLICATION

Human cloning could be done tomorrow.
—Alan Trounson, in vitro fertilization clinician, Monash University

Even if we had to transfer the laboratory on a boat located in international waters, the human cloning project will continue.
—Rael, ex-race car driver and founder of Clonaid company

Thus, the postmodern adventure of the reconstruction of nature begins with the genetic engineering of transgenic animals and the cloning of numerous animal species for agricultural, medical, and scientific purposes— while in fact biotechnology is being positioned as a field for prodigious profits. The fate of the human is inseparable from the future of our fellow animal species: They are the launch pad for the redesign of human nature. With the birth of Dolly, a new wave of animal exploitation arrived, and anxiety grew about a world of cloned humans that scientists said was technically feasible and perhaps inevitable. Ian Wilmut, head of the Roslin Institute team that cloned Dolly, is an example of an animal and stem cell cloning advocate who repudiates human replication. Like Jaenisch and numerous others, Wilmut believes human cloning is unethical, unnecessary, and dangerous, and that the inevitable deformities would be cruel to both the parents and children involved (Wilmut et al., 2000).

Wilmut feels human cloning should not be attempted until there is a quantum leap in cloning technologies—an advance he feels is at least 50 years away. Most of all, Wilmut fears that the drive toward human cloning could cause a backlash against all cloning, and thereby thwart the far more important research into cloning stem cells for therapeutic purposes. For Wilmut, the authentic purpose for biotechnology is to cure disease and improve agriculture. Whatever his intention, however, many scientists and entrepreneurs inspired by the Roslin Institute's work have aggressively pursued the goal of human cloning as the true telos of genomic science. Driven by market demands for clones of infertile people, of those who have lost loved ones, of gays and lesbians who want their own children, of those who want to clone themselves or family members to provide needed organs, and of numerous other client categories, doctors and firms are actively pursuing human cloning.

The Race to Clone Humans

Pro-human cloning forces include Richard Seed, who shocked the world in 1997 by declaring that he was prepared to clone himself, later appending the project to his wife. The Raelins, a wealthy Quebec-based religious cult, believe that all humans were cloned in laboratories by alien scientists and claim that their Clonaid project is about to produce the first human clone (which they initially projected to be ready by November 2001). Infertility specialists Severino Antinori and Panayiotis Zanos openly announce their intent to clone humans in defiance of any national law if necessary. The Council for Secular Humanism is a broad coalition of scientists, philosophers, authors, and politicians who decry the influence of religion in the cloning debates and champion the cause of human cloning as they assure us that cloning will not create any "moral predicaments beyond the capacity of human reason to resolve."[24] The Human Cloning Foundation is an internet umbrella group for diverse clonistas who see cloning as the best hope for curing infertility and diseases and promoting longevity.[25]

One bioethicist estimates that there are currently at least a half dozen laboratories around the world doing human cloning experiments.[26] Although cloning human beings is illegal in the United States, Britain, Germany, Japan, and elsewhere, in other places (e.g., Asia, India, Russia, and Brazil), it is perfectly legal, and human cloning is being pursued both openly and clandestinely. In fact there are at least two known cases where human embryos have been cloned, but the experiment was terminated. According to *Wired,*

> In 1988, a scientist working at Advanced Cell Technology in Worcester, Massachusetts took a human somatic cell, inserted it into an enucleated cow egg, and started the cell dividing to prove that oocytes from other species could be used to create human stem cells. He voluntarily stopped the experiment after several cell divisions. A team at Kyung Hee University in South Korea said it created an embryonic adult human clone in 1999 before halting the experiment, though some doubt that any of this really happened. Had either of these embryos been placed in a surrogate mother, we might have seen the first human clone. (9.02, February 2001, p. 128)

In November 2001, ACT created a global sensation with (misleading) reports they had cloned human embryos. Although many scientists think human cloning is possible and inevitable, some think it is likely that human clones already exist, perhaps in hideous form where they are studied on an island, such as was portrayed in H. G. Wells' *The Island of Dr. Moreau* (Best & Kellner, 2001). The breeding of monstrosities in animal cloning, the pain and suffering produced, and the possibility of assembly production of animals and humans should give pause to those who want to plunge ahead with human cloning. Animal cloning experiments produced scores of abnormalities, and it is highly likely that human cloning would do the same.[27]

The possibilities of producing serious human defects raises ethical dilemmas as well as the question of the social responsibility involved in the care of deformed beings produced by human cloning experiments. Fervent pro-cloners like Antinori and Zavos deny there are any risks to cloning humans and claim that there is enough information to proceed with confidence. If pressed to admit there might be mistakes, they simply write them off as necessary means to the end of reproductive freedom and medical progress. Ignoring the availability of frozen embryos and existing children for adoption, they claim the right to reproduce as crucial for human beings, and argue that this right, which in fact does not exist in any social constitution, outweighs any risks to the baby or society as a whole once the doorway is opened to the world of human cloning.

At present, however, what sane person would want to produce a possibly freakish replication of him or herself or a dead loved one? What are the potential health risks to women who would be called on to give birth to human clones, at least before artificial wombs make women, like men, superfluous to the reproductive process? Who will be responsible for caring for deformed human clones that parents renounce? Is this really an experiment that the human species wants to undertake so that self-centered infertile couples can have their own children (apparently some can only love a child with their own DNA) or misinformed narcissists can spawn what they think will be their carbon-copy twins? What happens if human clones breed? What mutations could follow? What might result from long-range tampering with the human genome as a consequence from genetic engineering and cloning?

Furthermore, until scientists figure out how to clone minds, cloning inevitably involves reproduction of bodily DNA, raising questions of what sorts of minds cloning might produce. What if cloned humans appear to be mentally defective or aberrant as a result of the technology? What might be the long-term costs of the perceived short-term benefits that cloning may produce? Already scientists are raising the issue of cognitive deficiencies in cloned animals, and certainly this problem is relevant to the project of human cloning.

In addition, as the TV series "Dark Angel" illustrates, there is the possibility of a military appropriation of cloning to develop herds of *Übermenschen* (although no two would be exactly alike). Indeed will commodification of the humane genome, eugenics, designer babies, and genetic discrimination all follow as unavoidable consequences of helping infertile couples and other groups reproduce, or will human cloning become as safe and accepted as in vitro fertilization (IVF), once also a risky and demonized technology? Will developing countries be used as breeding farms for animals and people, constituting another form of global exploitation of the have-nots by the haves? What are the consequences of the commodification of the human genome and the patenting of stem cells and research methods?

With so many questions and uncertainties that arise, it is clear that the project of human cloning is being approached in a purely instrumental and mechanistic framework that does not consider long-term consequences to the human genome, social relations, or ecology. If social relations and consequences are considered, likely this is from the perspective of improving the Nordic stock and creating an even deeper cleavage between rich and poor because, without question, only the rich will be able to afford genetically designed and/or cloned babies with superior characteristics. This situation could change if the state sponsors cloning welfare programs or the prices of a "Gen-Rich" (Silver, 1998) baby drop like computers, but the wealthy will already have gained a decisive advantage, and "democratic cloning" agendas beg the question of the soundness of human cloning in the first place.

Problems With Human Cloning

Thus, we have serious worries about biotechnology not only due to the colonialist history of science and capitalism, the commodification of the life sciences, and how genetic technologies have already been abused for profit and power by corporations like Monsanto and Du Pont, but also because of the reductionistic paradigm informing molecular engineering.[28] Ironically, although biology helped shape a postmodern physics through evolutionary and holistic emphases, the most advanced modes of biological science—genetic engineering and cloning research—have not advanced to the path of holism and complexity (Best & Kellner, 2001). Rather, biotechnology seems to have regressed to the antiquated errors of atomism, mechanism, determinism, and reductionism. The new technosciences and the outmoded paradigms (Cartesian) and domineering mentalities (Baconian) that inform them generate a volatile mix, and the situation is gravely exacerbated by the commercial imperatives driving research and development—the frenzied gene rush toward DNA patenting.

Yet if human cloning technologies follow the path of IVF technologies, eventually they will become widely accepted, although currently large percentages of U.S. citizens oppose it (90% according to some polls in summer 2001). Alarmingly, scientists and infertility clinics have taken up human cloning technologies all too quickly. After the announcement of the birth of Dolly, many were tripping over themselves to announce emphatically that they would never pursue human cloning. Nonetheless, only months later, these same voices began to embrace the project.[29] The demand from people desperate to have babies, or resurrect their loved ones in conjunction with the massive profits waiting to be made, is too great an allure for corporations to resist—a demand begging for supply. The opportunistic attitude of cloning advocate Panayiotis Zavos is all too typical: "Ethics is a wonderful word, but we need to look beyond the ethical issues here. It's not an ethical issue [!]. It's a medical issue. We have a duty here. Some people need this to complete the life cycle, to reproduce."[30]

In his attempt to dispel the ineliminable moral quandaries surrounding cloning, Zavos has confused need with desire and reduced humans to crude reproduction machines. Yet as his statement shows, defenders of cloning and biotechnology argue for the primacy of individual reproductive rights over potential risks to society as a whole. They believe that science is valuable to the extent that it increases freedom, individuality, and choice, as if embryos were a soft drink and what an individual chooses in this case is not of enormous consequence for future humanity, to say nothing of the deformed children who surely will be the guinea pigs of science. Of them Zavos can only say, "We're ready to face those mishaps. . . . It's part of any price that we pay when we develop new technology."[31]

There are indeed legitimate grounds for anxiety and loathing of cloning, but most fears of human cloning are irrationally rooted in what Leon Kass claimed is an intuitive human repulsion—the "yuk" factor—toward something that is seemingly unnatural (Kass & Wilson, 1998; critiqued by Pence, 1998). Many such clonophobic arguments are weak. The standard psychological objections, in particular, are poorly grounded. We need not fear Hitler armies assembling because the presumption of this dystopia—genetic determinism—is false (although certain desirable traits could be cloned that might prove useful for military powers). Nor need we fear individuals unable to cope with lack of their own identity because identical twins are able to differentiate themselves from one another relatively well and they are even more genetically similar than clones would be. Nor would society always see cloned humans as freaks because people no longer consider testtube babies alien oddities, and there are anywhere from 20,000 to 200,000 such humans existing today (figures vary widely). The physiological and psychological dangers are real, but in time cloning techniques could be perfected so that cloning might be as safe as, as if not safer than, babies born through a genetic throw of the dice—or IVF.

A strong objection against human cloning and genetic engineering technologies is that they could be combined to design and mass reproduce desirable traits, bringing about a society organized around rigid social hierarchies and genetic discrimination—as vividly portrayed in the film *Gattaca* (1997). This was of course the nightmare of Aldous Huxley, who continued H. G. Wells' speculations on a genetically engineered society and creation of a new species. Indeed with only trivial qualifications, Huxley's (1989a [1932]) *Brave New World* of genetic engineering, cloning, addictive pleasure drugs (soma), entertainment and media spectacles, and intense social engineering has arrived. Huxley thought cloning and genetic engineering were centuries away from realization, but in fact they began to unfold a mere two decades since his writing of *Brave New World* in the early 1930s. For instance, technocapitalism cannot yet biologically clone human beings, but it can clone them in a far more effective way—socially. Whereas biological clones would have a mind of their own, because the social world and experiences that conditioned the original could not be reproduced, social cloning according to a given ideological and functional model is far more controlling. That is why Huxley's (1989b [1958]) sequel work, *Brave New World Revisited*, focused on various modes of social conditioning and mind control.

Defenders of cloning and biotechnology argue that current science is geared toward increasing individuality and choice, enabling people to design their own children and within limits to mold their own bodies. Already parents can genetically choose the sex of the their child. Soon they might be able to isolate and remove genes that cause obesity, addictions, and a host of fatal illness, as well as engineer genes that would enhance intelligence, strength, athleticism, physical attractiveness, and other desirable traits.

Of course as Baudrillard (2000) argued, cloning is connected as well to the fantasy of immortality, to defeating the life–death cycle. Techno-utopians fantasize about the possibility of cloning one's body or downloading one's memories into another body or machine, thereby achieving immortality and alleged continuity of selfhood. The Raelians promote cloning as a chance for eternal life. In the current social setting, it is no surprise that cryogenics—the freezing of dead human beings in the hope that they might be regenerated in the future through medical advances—is a booming global industry.

Currently, the human race stands at a crossroads and must make crucial choices concerning the future of the human, including the issue of cloning. Whatever one's philosophical and ethical conceptions of cloning, it is clear that at present human cloning is unacceptable. Proponents of human cloning argue that it took hundreds of attempts to develop a testtube baby and that trial and error is simply the scientific method. We need to ask, however, whether such costs are legitimate when the benefits are not yet clear. Although one might sympathize with couples that fervently desire a child and utilize IVF, legions of unwanted children await adoption, and it is difficult to justify the great leap forward to cloning through these kinds of rationales.

Therapeutic Versus Reproductive Cloning: The Debate Over Stem Cell Research

> It is not unrealistic to say that stem cell research has the potential to revolutionize the practice of medicine.
> —Dr. Harold Varmus, former NIH director

> The 20th century was the drug therapy era. The 21st century will be the cell therapy era.
> —George Daleuy, biologist with the Whitehead Institute for Biomedical Research, Cambridge, Massachusetts

Full-blown human reproductive cloning is problematic for numerous reasons, and we reject it on the grounds that it lacks justification and portends a world of eugenics and genetic discrimination rooted in the creation and replication of desired human types. Yet scientists are also developing a more benign and promising technology of stem cell research or therapeutic cloning. The controversy around embryonic stem cell research—because it involves using and destroying cells from frozen human embryos—remains one of the key debates of our time, important enough to provoke a major policy crisis for the Bush administration and warrant an address to the nation on prime-time TV in August 2001. Rarely do scientific debates erupt into the public forum. Although the technical aspects are difficult and complex, the ethical and medical stakes are clear enough to command a national debate.

In 1998, Dr. James A. Thomson, a developmental biologist at the University of Wisconsin, announced to the scientific world that he had isolated embryonic stem cells, thus portending a new era of regenerative medicine based on the renewal and re-creation of the body's cells. Stem cells are the primitive master cells of the body that differentiate into functions like skin, bone, nerve, and brain cells (the body produces over 200 cell types). The goal of stem cell research is to program the development of stem cells toward specific functions to replace lost or damaged cells, tissues, and organs. Using similar technological breakthroughs such as led to Dolly, stem cell research involves cloning cells from a wide range of human tissue or young human embryos (around 5 days of age) and aborted fetal tissues.

In the debates over stem cell research, an important distinction emerged between adult stem cells, which are derived from blood, bone marrow, fat, and other tissues, and embryonic stem cells from discarded IVF cultures, aborted fetuses, or embryos created in a lab. Although scientists are experimenting with adult stem cells, the current consensus is that embryonic cells are the most pliable and hence have the most regenerative potential. In July 2001, the National Institutes of Health (NIH) issued a report that, "Stem cells from adults and embryos both show enormous promise for treating an array of diseases but at this early stage, cells from days-old embryos appear to offer certain key advantages." As Ceci Connolly summarized it: "Embryonic stem cells are more plentiful and therefore easier to extract, can be grown and made to multiply in the laboratory more easily and appear to have the uncanny ability to develop into a much wider array of tissues."[32] In fact embryonic and adult stem cell research may each contribute to significant medical and health advancements. According to Senator Bill Frist (R–Tenn), the only medical doctor in Congress, an opponent of abortion, and key science advisor to the Bush administration: "Because both embryonic and adult stem cell research may contribute to significant medical and health advancement, research on both should be federally funded within a carefully regulated, fully transparent framework that ensures respect for the moral significance of the human embryo."[33]

Scientists argue that therapeutic cloning has tremendous medical potential. For example, early in life, each individual could have his or her stem cells frozen to create a "body repair kit" if she or he ever developed a disease or even lost a limb. There would be no organ shortages, no rejection problem, and no need for animal exploitation because the cells would be their own. Although as of yet there have been no significant advances in human research, and the results so far confined to animals are not necessarily applicable to human beings, stem cell research nonetheless shows remarkable potential for revolutionary breakthroughs in medicine. Among their achievements with mice, rats, pigs, and fetal monkeys, scientists have directed stem cells to (a) produce insulin, (b) induce growth of brain cells, and (c) form new blood vessels in hearts, thereby suggesting immense contributions to curing diabetes, Alzheimer's or Parkinson's, and heart disease.[34] Still although industries and media often hype the research as producing imminent medical revolutions, many scientists believe breakthroughs in gene therapy and therapeutic cloning are likely decades away and expectations have been unduly raised.[35]

Another crucial distinction involves using embryonic stem cells from IVF discards and cloning embryos for the explicit sake of research. Whereas Britain allows both kinds of stem cell research, and thus condones embryo cloning for therapeutic purposes, the Bush administration highly restricts the use of IVF stem cell lines and condemns embryonic cloning. Yet many scientists argue that the ideal source of stem cells for regenerative medicine would not only be those derived from IVF embryos, but from embryos cloned from a patient's own cells because the derived stem cells would be one's own and in theory far less susceptible to rejection. Thus, there is a medical justification for cloning human embryos, and embryo cloning will be crucial to regenerative medicine.

Many religious groups and hard-core technology critics vituperate against stem cell research as violating the inherent sanctity of life. To be sure, there is an ethical issue at stake in creating embryos for research purposes or even using IVF cells because living matter is being used as a means to some end other than its own existence. Clearly, using IVF cells that are going to be destroyed is less objectionable than cloning an embryo for the sake of harvesting its cells and then terminating it, but many religious groups and conservatives nonetheless vehemently oppose all forms of stem cell research and any manipulation of life no matter what profound medical consequences may result. "Anyone truly serious about preventing reproductive human cloning must seek to stop the process from the beginning," Leon Kass, later to be Bush's cloning czar, proclaimed before a House judiciary subcommittee in June 2001.[36]

To challenge stem cell research, many conservatives (and some liberals) are recycling philosophical arguments from earlier debates over abortion. The Pope and critics of stem cell research argue that once a sperm and egg are mixed into an embryo, no matter what the medium, there is a human life with all of its rights and sacredness. Others claim that a human life exists only when the embryo is implanted in a mother and has undergone the beginnings of the maturation process. Some medical experts assert that 14 days is the crucial dividing line when a backbone and organs begin to develop, whereas many prochoice proponents argue that a fetus is not yet fully a human being. These earlier philosophical arguments have been revived in the stem cell debate to legitimate conflicting scientific and political positions.[37] In the context of stem cell research, religious conservatives recycle the same question-begging argument: (a) a human embryo is a human being, (b) it is wrong to take a human life, and (c) therefore, it is wrong to destroy an embryo. The most controversial claim of the argument, in premise (a), is either just assumed or defended through dogmatic claims that "life begins at conception," when arguably there is no real conception in a petri dish holding a 5-day-old cell mass.[38]

Ultimately, the debate comes down to the philosophical issue of what constitutes a human being. Opponents of therapeutic human cloning and embryonic stem cell research claim that conception takes place when an embryo is produced even in a petri dish. Critics of this notion of human life argue that an embryo is a merger of sperm and egg that takes place in 5 or 6 days and is called a blastocyst, which scientists distinguish from a fetus. Scientists further claim that an embryo only attains fetus status at around 14 days, when it develops a primitive streak—the beginnings of a backbone. Up until that point, a single embryo can divide into identical twins, and two embryos can merge into one, leading Ronald Green, a Dartmouth bioethicist, to conclude: "It is very clear that you cannot speak of a human individual in the first 14 days of development. How can one speak of the presence of an individual soul if the embryo can be split into two or three?"[39]

Clearly it is difficult to say when human life begins, and claims that it emerges at conception are simplistic. So far human life has only been produced from fetuses that mature in the womb of a woman's body, and thus we have trouble conceiving that 5-day-old embryos in a petri dish are human. It also might be pointed out that only about one in eight embryos implanted through IVF achieves fetal status, and few conservative critics worry over the doomed embryos or question the ethics of IVF as a whole—a technology that produces surplus cells for medical research. The fact that embryos typically used for stem cell research are leftover from couples using in vitro fertilization and are marked for destruction regardless strongly undercuts the force of the argument against embryonic stem cells.[40]

Indeed the slippery slope argument beloved by conservatives (the direct and unavoidable path from stem cell research to fetus farms and a society peopled by clones) is easily turned against them. In the age of cloning where possibly any cell can be replicated and turned into an embryo, one might argue that it is unethical even to scrape any skin cells because they too are potential human beings.[41] Silly, perhaps, but this is also an indicator of the surreality of the postmodern adventure. In an amazing alchemy, scientists can directly transform cells of one kind into another. PPL Therapeutics succeeded in transforming a cow's skin cell into a basic stem cell and then refashioned it as a heart cell. Further, researchers are working on cultivating spermless embryos, studying how to prod unfertilized eggs to grow to produce stem cells.[42] Geron has created heart cells that beat in a petri dish. Clearly the implications of stem cell research are staggering.

One should not see the use or creation of human embryos for medical resources as a trivial issue, but the debate over therapeutic cloning involves competing values and conceptions of the nature of a human being. This is a conflict between a small clump of cells no bigger than the period at the end of this sentence and full-fledged human beings in dire medical need. In a conflict between a tiny ball of nonsentient cells or fetuses that would be disposed of regardless, and full-fledged human beings suffering from diseases that lack a cure, most people would choose the latter category of human persons.

Thus, although many conservatives defend the sanctity of embryonic cells, and so far are successfully thwarting stem cell research, thousands of people continue to suffer and die from Alzheimer's, Parkinson's disease, diabetes, paralysis, and other afflictions. This is a strange position for prolife and compassionate conservatives to defend. The entire moral quandary may be blunted, however, because scientists are now discovering ways to use stem cells derived from umbilical cords, bone marrow, and even fat cells.[43]

DEFERRING THE BRAVE NEW WORLD: CHALLENGES FOR DEMOCRACY

> *Cloning is inefficient in all species. Expect the same outcome in humans as in other species: late abortions, dead children and surviving but abnormal children. . . .*
> —Ian Wilmut

> *Is there any risk too great or any reason too trivial for you not to attempt human cloning?*
> —Alta Charo, University of Wisconsin bioethicist,
> speaking to Antinori and Zavos

Thus, by summer 2001, a technical and esoteric debate over stem cells, confined within the scientific community during the past years, had moved to the headlines to become the forefront of the ongoing science wars—battles over the cultural and political interpretations and implications of science (Best & Kellner, 1997). The scientific debate over stem cell research in large part is a disguised culture war, and conservatives, liberals, and radicals have all jumped into the fray. For example, in our own case, coming from a perspective of critical theory and radical democratic politics, we reject conservative theologies and argue against conflations of religion and the state. Likewise, we question neoliberal acceptance of corporate capitalism and underscore the implications of the privatization of research and the monopolization of knowledge and patents by huge biotech corporations. In addition, we urge a deeper level of public participation in science debates than do conservatives or liberals, and we believe that the public can be adequately educated to have meaningful and intelligent input into technical issues such as cloning and stem cell research.

As we have shown, numerous issues are at stake in the debate over cloning—having to do not only with science, but also with religion, politics, economics, democracy, and the meaning and nature of human beings and all life forms as they undergo a process of genetic reconstruction. Thus, our goal throughout this chapter has been to question the validity of the cloning project, particularly within the context of a global capitalist economy and its profit imperative, a modernist paradigm of reductionism, and a Western sensibility organized around the concept of the domination of nature. Until science is recontextualized within a new holistic paradigm informed by a respect for living processes, by democratic decision making, and by a new ethic toward nature, the genetic sciences on the whole are in the hands of those governed by the imperatives of profit. Moreover, they are regulated by politicians who do not have a good grasp of the momentous issues involved, requiring those interested in democratic politics and progressive social change to educate and involve themselves in the politics of biotechnology.

We have already entered a new stage of the postmodern adventure in which animal cloning is highly advanced and human cloning is on the horizon, if not now underway. Perhaps little human clones are already emerging, with failures being discarded, as were the reportedly hundreds of botched attempts to create Louise Brown, the first test-tube baby, in 1978. At this stage, human cloning is indefensible in light of the possibility of monstrosities, dangers to the mother, burdens to society, failure to reach a consensus on the viability and desirability of cloning humans, and the lack of compelling reasons to warrant this fateful move. The case is much different, however, for therapeutic cloning, which is incredibly promising and offers new hope for curing numerous debilitating diseases. But even stem cell research, and the cloning of human embryos, as we have seen are problematic in part because they are the logical first step toward reproductive cloning and mass production of desired types, which unavoidably brings about new (genetic) hierarchies and modes of discrimination.

Thus, we need to discuss the numerous issues involved in the shift to a posthuman, postbiological mode of existence where the boundaries between our bodies and technologies begin to erode as we morph toward a cyborg state. Our technologies are no longer extensions of our bodies, as

Marshall McLuhan stated, but rather are intimately merging with our bodies as we implode with other species through the genetic crossings of transgenic species. In an era of rapid flux, our genotypes, phenotypes, and identities are all mutating. Under the pressure of new philosophies and technological change, the humanist mode of understanding the self as a centered, rational subject has transformed into new paradigms of communication and intersubjectivity (Habermas, 1984/1987, 1991) and information and cybernetics (Hayles, 1999).

Despite these shifts, it is imperative that elements of the modern Enlightenment tradition be retained as it is simultaneously radicalized. Now more than ever, as science embarks on the incredible project of manipulating atoms and genes through nanotechnology, genetic engineering, and cloning, its awesome powers must be measured and tempered through ethical, ecological, and democratic norms in a process of public debate and participation. The walls between experts and laypeople must be broken down along with the elitist norms that form their foundation. Scientists need to enter dialogical relations with the public to discuss the complexities of cloning and stem cell research, to make their positions clear and accessible, accountable and responsible, while public intellectuals and activists need to become educated in biotechnology to engage in debate in the media or public forums on the topics.

Scientists should recognize that their endeavors embody specific biases and value choices, subject them to critical scrutiny, and seek more humane, life-enhancing, and democratic values to guide their work. Respect for nature and life, preserving the natural environment, humane treatment of animals, and serving human needs should be primary values embedded in science. When these values conflict, as in the tension between the inherent value of animals and human needs, the problem must be addressed as sensitively as possible.

This approach is quite unlike how science so far has conducted itself in many areas. Most blatantly, perhaps, scientists, hand in hand with corporations, have prematurely rushed the genetic manipulation of agriculture, animals, and the world's food supply while ignoring important environmental, health, and ethical concerns. Immense power brings enormous responsibility, and it is time for scientists to awaken to this fact and make public accountability integral to their ethos and research. A schizoid modern science that rigidly splits facts from values must give way to a postmodern metascience that grounds the production of knowledge in a social context of dialogue and communication with citizens. The shift from a cold and detached neutrality to a participatory understanding of life that deconstructs the modern subject–object dichotomy derails realist claims to unmediated access to the world and opens the door to an empathetic and ecological understanding of nature (Birke & Hubbard, 1995; Keller, 1983).

In addition, scientists need to take up the issue of democratic account-ability and ethical responsibility in their work. As Bill Joy (2000) argued in a much-discussed *Wired* article, uncontrolled genetic technology, artificial in-telligence, and nanotechnology could create catastrophic disasters as well as utopian benefits. Joy's article set off a firestorm of controversy, espe-cially his call for government regulation of new technology and relinquish-ment of development of potentially dangerous new technologies, as he claimed biologists called for in the early days of genetic engineering, when the consequences of the technology were not yet clear.[44] Arguing that scien-tists must assume responsibility for their productions, Joy warned that hu-mans should be very careful about the technologies they develop because they may have unforeseen consequences. Joy noted that robotics was pro-ducing increasingly intelligent machines that might generate creative ro-bots that could be superior to humans, produce copies of themselves, and assume control of the design and future of humans. Likewise genetic engi-neering could create new species, some perhaps dangerous to humans and nature, whereas nanotechnology might build horrific "engines of destruc-tion" as well as of the "engines of creation" envisioned by Eric Drexler.

Science and technology, however, not only require responsibility and ac-countability on the part of scientists, but also regulation by government and democratic debate and participation by the public. Publics need to agree on rules and regulations for cloning and stem cell research, and there need to be laws, guidelines, and regulatory agencies open to public input and scrutiny. To be rational and informed, citizens need to be educated about the complexities of genetic engineering and cloning—a process that can unfold through vehicles such as public forums, teach-ins, and creative use of the broadcast media and internet.

An intellectual revolution is needed to remedy the deficiencies in the education of both scientists and citizens, such that each can have, in Habermas' (1984/1987) framework, "communicative competency" in-formed by sound value thinking, skills in reasoning, and democratic sensi-bilities. Critical and self-reflexive scrutiny of scientific means, ends, and procedures should be a crucial part of the enterprise. "Critical," in Hara-way's (1997) analysis, signifies "evaluative, public, multiactor, multi-agenda, oriented to equality and heterogeneous well-being" (p. 95). Indeed there should be debates concerning precisely what values are incorpo-rated into specific scientific projects and whether these serve legitimate ends and goals. In the case of mapping the human genome, for instance, enormous amounts of money and energy are being spent, but almost no resources are going toward educating the public about the ethical implica-tions of having a genome map. The Human Genome Project spent only 3% to 5% of its $3 billion budget on legal, ethical, and social issues, and Celera spent even less.[45]

A democratic biopolitics and reconstruction of education would involve the emergence of new perspectives, understandings, sensibilities, values, and paradigms that put in question the assumptions, methods, values, and interpretations of modern sciences, calling for a reconstruction of science (on "new science" and "new sensibilities"; Marcuse, 1964, 1969). At the same time, as science and technology co-construct each other, and both coevolve in conjunction with capitalist growth, profit, and power imperatives, science is reconstructing—not always for the better—the natural and social worlds as well as our identities and bodies. There is considerable ambiguity and tension in how science will play out given the different trajectories it can take. Unlike the salvationist promises of the techoscientific ideology and the apocalyptic dystopias of some of its critics, we see the future of science and technology to be entirely ambiguous, contested, and open. For now the only certainty is that the juggernaut of the genetic revolution—as that of digital information technologies—is rapidly advancing and that in the name of medical progress animals are being victimized and exploited in new ways while the replication of human beings is looming.

The human race is thus at a terribly difficult and complex crossroads. Whatever steps we take, it is imperative that we not leave the decisions to the scientists anymore than we would to the theologians (or corporate-hired bioethicists for that matter) because their judgment and objectivity is less than perfect, especially for the majority who are employed by biotechnology corporations and have a vested interest in the hastening and patenting of the brave new world of biotechnology.[46] The issues involving genetics are so important that scientific, political, and moral debate must take place squarely within the public sphere. The fate of human beings, animals, and nature hangs in the balance, thus it is imperative that the public become informed on the latest developments and biotechnology and that lively and substantive democratic debate take place concerning the crucial issues raised by the new technosciences.

ENDNOTES

1. This chapter draws on work from our 2001 book, *The Postmodern Adventure*, and is part of a larger project we are developing on cloning and stem cell research. Thanks to Sandra Braman for helpful comments on several drafts of our chapter.

2. The much-cited figure of "277 failed attempts" at producing Dolly is often used indiscriminately to conjure up a Gothic image of a dungeon of monstrosities. As Silver (1998) explained, "The implication—sometimes stated explicitly—was that many lambs died or were born with genetic malformations. What [the number 277] stood for was the number of fusions that were initially obtained between donor cells and unfertilized eggs. Only 29 of these fused cells actually became embryos, and these 29 embryos were introduced into 13 ewes, of which one became pregnant and gave birth to Dolly" (p. 120). Nonetheless, as we argue, cloning as a whole is plagued with failures and deformed or sickly creations. See Mi-

chael Woods, "Deaths, birth defects hover over cloning process," *Toledo Blade* (Aug. 15, 2001), and Gina Kolata, "In cloning, failure far exceeds success," *The New York Times*, Dec. 11, 2001.

3. Cited in Carey Goldberg and Gina Kolata, "Scientists announce births of cows cloned in new way," *The New York Times*, January 21, 1998, p. A14. Companies are now preparing to sell milk from cloned cows; see Jennifer Mitol, "Got cloned milk?" abcnews.com, July 16, 2001. For the story of Dolly and animal cloning, see Kolata (1998).

4. See Sheryl Gay Stolberg, "Breakthrough in pig cloning could aide organ transplants," *The New York Times*, Jan. 4, 2001.

5. See http://abcnews.go.com/sections/DailyNews/biotechgoats. 000618.html.

6. See Heather Moore, "The modern-day island of Dr. Moreau," http://www.alternet.org/story.html?StoryID=11703, October 12, 2001. For a vivid description of the horrors of animal experimentation, see Singer (1975); for an acute diagnosis of the unscientific nature of vivisection, see Greek and Greek (2000).

7. See our discussion of Wells in Best and Kellner (2001).

8. See "In test, Japanese have no beef with cloned beef," http://www.washingtonpost.com/wpsrv/inatl/daily/sept99/japan10.htm. According to one report, it is more accurate to refer to this beef as being produced by "embryo twinning" and not the kind of cloning process that produced Dolly; see "Cloned beef scare lacks meat," http://www.wired.com/news/technology/0,1282,19146,00.html. As just one indicator of the corporate will to clone animals for mass consumption, the National Institute of Science and Technology has donated $4.7 million to two industries to fund research into cloning chickens for food. See "Cloned chickens on the menu," New Scientist.com, August 15, 2001.

9. See Heather Moore, *op cit.*, and Sharon Schmickle, "It's what's for dinner: Milk and meat from clones," www.startribune.com/stories/462/868271.html, December 2, 2001.

10. "Clonefarm: Billions of identical chickens could soon be rolling off production lines," www.newscientist.com/hottopics/cloning/cloning.jsp?id=23040300, August 18, 2001.

11. Cited in Heather Moore, *op cit.*

12. "The Missyplicity Project boasts a strong code of bioethics"; see http://www.missyplicity.com/.

13. See http://www.transgenicpets.com/.

14. Gina Kolata, "In cloning, failure far exceeds success," www.nytimes.com/2001/12/11/science/11CLON.html.

15. See "Clones contain hidden DNA damage," www.newscientist.com/news/news.jsp?id=ns9999982; see also the study published in *Science* (July 6, 2001), which discusses why so many clone pregnancies fail and why some cloned animals suffer strange maladies in their hearts, joints, and immune system.

16. "Clone study casts doubt in stem cells: Variations in mice raise human research issues," www.washingtonpost.com/ac2/wp-dyn/A23967-2001Jul5?language=printer, July 6, 2001.

17. See "Scientists warn of dangers of human cloning," www.abcnews.com. See also the commentaries in Gareth Cook, "Scientists say cloning may lead to long-term ills," *The Boston Globe*, July 6, 2001; Steve Connor, "Human cloning will never be safe," *Independent*, July 6, 2001; Carolyn Abraham, "Clone creatures carry genetic glitches," July 6, 2001. Connor cites Dolly-cloner Ian Wilmut who noted: "It surely adds yet more evidence that there should be a moratorium against copying people. How can anybody take the risk of cloning a baby when its outcome is so unpredictable?"

18. "Report says scientists see cloning problems," http://abcnews.go.com/wire.US/reuters20010325 573.html.

19. The Westhusin quote is at abcnews.go.com/cloningflaw010705. htm; the "misplaced carbons" quote is in Philip Cohen, Clone killer, www.newscientist.com/news.

20. The fact that science regards so much of the human genome as *junk DNA* is grounds enough for being suspicious of an arrogant attitude that claims science can distinguish between what is essential and inessential, when in fact the inessential or junk DNA may have important functions of which science is unaware. Similarly, although negative eugenics seems unqualifiably promising in its bid to rid the genome of bad genes that cause certain diseases, it is not clear what the impact may be of deleting such genes from the human genome through something like germ-line therapy, which makes permanent alterations in the genetic code.

21. "Human clone moves sparks global outrage," www.smh.com.au, March 11, 2001.

22. Given this attitude, it is no surprise that in September 2001, Texas A&M University, the same institution working on cloning cats and dogs, showed off newly cloned pigs, who joined the bulls and goat already cloned by the school, as part of the "world's first cloned animal fair."

23. Pamela Weintraub, "Back from the brink: Cloning endangered species," http://news.bmn. com/hmsbeagle/109/notes/feature2, August 31, 2001; "Gene find no small fetus," www.wired. com/news/print/0,1294,41513,00.html.

24. "Declaration in defense of cloning and the integrity of scientific research," www. secularhumanism.org/library/fi/cloning declaration 17 3.html.

25. See http://www.humancloning.org/.

26. Investigative reporter Joe Lauria found a secret cloning lab supposedly carrying out Raelian human cloning experiments, but it appeared abandoned, and there are suspicions that the whole effort was a fraud to exploit a desperate family that wanted its child cloned; see *London Times*, August 12, 2001. Arguments for human cloning are found at www. humancloning.org and www.reason.com/biclone.html. On predictions that human cloning experiments are already underway, see www.wired.com/wired/archive/9.02/projectxpr. html).

27. In August 2001, some scientists found that humans have two copies of a gene known to cause mutations in cloned nonprimate mammals, and so claimed that human cloning may actually be safer than animal cloning. Ian Wilmut, Lee Silver, and others, however, disputed this claim, arguing that the researchers misinterpreted their data and there are genes other than the one identified that can cause potential problems when expressed in a human clone. See "Study: Human cloning is safer," www.wired.com, August 2001.

28. See Harding (1998) for a discussion of how modern science and capitalism co-evolved in the context of colonialism, whereby they underpinned each other in the bid to control other peoples and exploit their knowledge.

29. See Gina Kolata, "Human cloning: Yesterday's never is today's why not?," *The New York Times*, December 2, 1997.

30. Cited in Nancy Gibbs, "Baby, "It's you! And you, and you . . . ," *Time*, February 19, 2001: 50. In March 2001, to great media fanfare, Zavos, Israeli biotechnologist Avi Bin Abraham, and Italian fertility specialist Severino Antinori announced that the group had signed up more than 600 infertile couples and were undertaking human cloning experiments to provide them with children; see "Forum on human cloning turns raucous," *Los Angeles Times*, March 10, 2001. When Zavos and his partner went to Israel to seek permission to do human cloning there, ABC News (March 25, 2001) reported that they received the blessing of a rabbi, but the Israeli justice minister said that he was against cloning "on moral and ideological grounds." A University of Pennsylvania ethicist said that Zavos had no medical training, had published no articles in the field, had no qualifications, and that one of the dangers of

cloning was that frauds were operating in the dangerous minefield of human cloning and exploiting people with false promises. There were also numerous discussions of the failures of animal cloning that were suggesting that human cloning would be highly dangerous and disturbing; see Aaron Zitner, "Perpetual pets, via cloning," *Los Angeles Times*, March 16, 2001; Gina Kolata, "Researchers find big risk of defect in cloning animals," *The New York Times*, March 25, 2001; and the examples provided next.

31. "Brave new world?" http://msnbc.com/news/525661.asp.

32. Ceci Connolly, "Embryo cells' promise cited in NIH study," *The Washington Post*, July 18, 2001, p. A01. The NIH notes the preliminary status of the report, the many uncertainties around stem cells, and the need for more research.

33. See www.time.com, July 19, 2001.

34. "Stem cells coaxed to produce insulin," http://www.msnbc.com/news/607294.asp; "Fetal stem cells boost brainpower," http://www.msnbc.com/news/566735.asp; "Rebuilding hearts," http://abcnews.go.com/sections/ GMA/DrJohnson/GMA010402Stemcells dr.Tim.html; and "Early success seen with 2nd type of stem cell," www.nytimes.com/2001/ 07/26/health/genetics/26MOUS.html. The experiment with brain cells involved injecting human stem cells from the brains of aborted fetuses into mice, rats, and pigs, thereby imploding species boundaries and demonstrating the versatility of human stem cells.

35. One key problem is that scientists as of yet have been unable to get stem cells to grow into the specialized types they seek, rather than clumps of different cells. For an important article that punctures much of the hype surrounding stem cell research, see "A thick line between theory and therapy, as shown with mice," Gina Kolata, www.nytimes.com/2001/12/18/ science/life/18MICE.html.

36. "Cloning Capsized?", *The Scientist*, *15*(16), p. 1, August 20, 2001.

37. The philosophical debate over when human life starts is a long-standing one. The Greek philosopher Aristotle chose 40 days into pregnancy, and the 40-day rule was long followed by Jewish and Muslim traditions. The Catholic Church followed this line until 1588, when Pope Sixtus V declared that contraception and abortion were mortal sins. However, the ruling was reversed 3 years later until 1859, when Pope Gregory XIV brought the church back to the view that the human embryo has a soul and renewed the call for excommunication for abortion at any stage. See Rick Weiss, "Changing conceptions," *The Washington Post*, July 15, 2001, p. B01.

38. For a thorough problematization of attempts to define the "beginning point" of life, see Silver (1998).

39. Cited in Aaron Zitner, "Uncertainty is thwarting stem cell researchers," *Los Angeles Times*, July 21, 2001, p. A01.

40. In Britain, the Human Fertilization and Embryology Authority has reported that some 50,000 babies have been born through in vitro fertilization since 1991, and 294,584 surplus human embryos have been destroyed. Although no official records have been kept in the United States, "According to the American Society for Reproductive Medicine, about 100,000 children have been born in the United States by in vitro fertilization, or twice the number in Britain, implying that some 600,000 embryos would have been destroyed if American clinics followed the same 5-year storage limit used in Britain. Only a small fraction of the discarded embryos would provide as many stem cells as researchers could use." See Nicholas Wade, "Stem cell issue causes debate over the exact moment life begins," *The New York Times*, August 15, 2001.

41. "Adult stem cells found in skin," www.newscientist.com/hottopics/cloning/cloning.jsp?id= ns99991147, August 13, 2001.

42. "Another advance for dolly cloners," www.wirednews.com/news/print/0.1294.41989.00.html; Aaron Zitner, "Working on sperm-less embryos," *Los Angeles Times*, August 12, 2001.

43. See "Adult approach to stem cells," http://www.wirednews.com/news/print/ 0,1294,38892,00.htm; "Need stem cells? It's in the fat," http://www.wired.com/news/print/ 0,1294,42957,00.html; "Human fat may provide useful cells," http://www.msnbc.com/news/ 557256.asp.

44. See the collection of responses to Joy's article in *Wired* 8.07 (July 2000). Agreeing with Joy that there need to be firm guidelines regulating nanotechnology, the Foresight Institute has written a set of guidelines for its development that take into account problems such as commercialization, unjust distribution of benefits, and potential dangers to the environment. See www.foresight.org/guidelines/current.html. We encourage such critical dialogue on both the benefits and dangers of new technologies and hope this chapter contributes to these debates.

45. See www.wired.com/news/0,1294,36886,00.html.

46. For a sharp critique of how bioethicists are coopted by corporations in their bid for legitimacy, see "Bioethicists fall under familiar scrutiny," http://www.nytimes.com/2001/08/02/ health/genetics/02BIOE.html).

9

Popular Representation and Postnormal Science: The Struggle Over Genetically Modified Foods

Graham Murdock
University of Loughborough

In the past year the United Kingdom has become the arena for a debate that will determine the kind of civilisation we fashion for ourselves in the twenty-first century—the debate over biotechnology and its commercial application . . . The United Kingdom has ignited a philosophical firestorm whose repercussions will be felt well into the next century.
—Rifkin, (1999, p. 1)

What we are witnessing is one of the greatest revolts against a new technology in history . . . it proposes a new relationship between politicians, corporations and consumers.
—Vidal (1999a, p. 20)

These statements, by a prominent campaigner in the United States and one of Britain's best-known environmental journalists, present the struggle over genetically modified foods in Britain during 1999 as a pivotal moment in the emerging politics around biotechnologies. This chapter sets out to map the course of events and tease out what they tell us about the shifting relations among commerce, science, communications, and popular mobilization.

Science solicits popular support on the basis that its discoveries will be applied in ways that make life steadily safer and more fulfilling for everyone. In recent years, however, this promise of cumulative progress has been eroded by a series of developments. These have disrupted the established assumptions and social relations of modern science and ushered in what has come to be called *postnormal* science (see Funtowicz & Ravetz, 1993). This is not an entirely satisfactory description. The prefix *post* overemphasizes the abruptness of the break with past conditions, whereas *normal* implies that relations between independent researchers and major power holders, and between scientific discoveries and technological innovations, proceed more smoothly than in fact they did (see Winston, 1998). Even so there is little doubt that the last two decades have seen major shifts.

First, research on global warming has demonstrated beyond a reasonable doubt that scientific interventions which originally appeared both limited and controllable can have unpredictable and far-reaching impacts the negative consequences of which may only become fully apparent after some decades. This new climate of risk and uncertainty has undermined science's strong association with progress and revivified popular conceptions of the scientist as a Frankenstein-like figure whose innovations may create more problems than they solve. Second, as the economic center of capitalism has moved from industrialized production to the manipulation of strategic information and meta-technologies, so command over core scientific knowledge has become a pivotal commercial resource prompting increasingly bitter skirmishes over who should own intellectual property and control the uses to which it is put.

The stakes in this struggle emerged particularly clearly in the clash between the entrepreneur, Craig Venter, and the university-based researcher, John Sulston, over the terms on which the results of the Human Genome Project would be released (see Sulston & Ferry, 2002). The data were generated by a transatlantic team of publicly funded scientists in which Sulston's laboratory, the Sanger Institute at Cambridge University, played a leading role. It offers a master key to understanding the genetic bases of individual capacities and vulnerabilities to disease. The commercial value, in terms of developing new medical and pharmaceutical products, is enormous. Venter wanted to exploit this potential by selling access to the data. Sulston and his collaborators insisted on making the full results publicly available. In the event, a compromise was reached. Sulston and his team released an open access draft of their results in 2001, and Venter marketed his own version based mainly on the results from the public project topped up with some data of his own.

The commercial value of new scientific knowledge coupled with the growing awareness of the dangers of applying it before adequate evaluations of possible environmental and health risks have been carried out and have moved questions of science to the center of the political arena where they are caught in the continual cross-fire between corporate interests and the common good. Commercial calculations emphasize short-term gains and leave the state to cope with any social or environmental damage that may emerge later. Citizens and consumers, in contrast, look to their children's future and expect the state to hold companies responsible for any long-term risks their interventions may cause.

This increasing politicization of science coincides with major shifts in the organization of both democratic politics and public communications. Over the last two decades or so, political parties and electoral systems have been steadily hollowed out as arenas for popular representation as popular engagement with contemporary issues, particularly among young people, has increasingly migrated to the various forms of activism organized by pressure groups and new social movements. These emerging political actors may not be able to match the conventional news management resources commanded by the established political parties and major corporations, but they have been quick to recognize the growing centrality of visual materials in popular news presentation and develop persuasive strategies based around image events (Delucca, 1999).

Earlier protest social movements perfected the "propaganda of the deed" launching newsworthy actions, such as planting a bomb or chaining oneself to the railings of a public building, in secret or without prior warning. The aftermath—a devastated building or a suffragette being led away by the police—might be caught on camera or followed up by a communique from the group, but often the action was left to speak for itself. Contemporary image events have to make their points in a news environment drenched in competing claims and photo opportunities. They cut through the clutter by mounting actions organized around strong central images that can distill public anxieties and detonate potent chains of association. These images are drawn overwhelmingly from the virtual visual archive established by popular fictions, comic books, films, and TV.

Scientists and government ministers tend to condemn the use of popular iconography in social debate, asserting that it oversimplifies complex issues, distorts the effective dissemination of scientific knowledge, and impedes rational discussion. This argument is oversimple, however. As anthropologists have long argued, far from contaminating communication, "habitual images and familiar metaphors provide the essential cultural forms that make ideas communicable" (Nelkin & Lindee, 1995, p. 12). As we will see, effective command of these symbolic resources played a central role in winning the contest of positions in the struggle over GM foods, and failure to grasp its importance played havoc with the best-laid plans.

FORTUNES AND REVERSALS

In 1998, almost three fourths of the 20.5 million hectares of land in the world devoted to cultivating genetically modified (GM) crops for commercial sale were to be found in the United States (Anderson, 1999). The major agrichemical companies that controlled the industry were eager to expand their operations by increasing their exports of GM foods and food ingredients and developing commercial plantings in other major economic zones. By selling new production technologies as well as novel products, leading companies in the sector, like Monsanto, hoped to maximize the returns on investments they had already made. Monsanto's profits at the time depended heavily on sales of its "Roundup Ready" soy beans, the seeds of which had been genetically engineered to resist the company's best-selling Roundup herbicide. By spraying fields before and during growing, farmers could eradicate weeds, leaving the modified soy beans unaffected. This locked farmers into an integrated production system that could not be unbundled and presented obvious gains for the company. Whether it benefited growers in terms of lower costs and higher yields was less clear cut, prompting increased concern about its implications for farmers' rights to choose their own production strategies.

The possible impact on the biodiversity of the natural ecosystems surrounding Roundup Ready planted fields and the possible risks to human health arising from the possibility that the soy plants might absorb some of the active ingredients in the herbicide, particularly Glyphosate, were also generating concern. Although there were no commercial plantings in the United Kingdom, GM soy and its derivatives were finding their way into the food chain in Britain through a range of common supermarket products.

The first consignments of U.S.-grown GM soy beans arrived in the United Kingdom in 1996 and contained unsegregated mixtures of GM and standard varieties, leaving consumers unable to avoid GM ingredients even if they so wished. Monsanto defended this practice, arguing that the two forms were "substantially equivalent" because comparisons of the chemical characteristics showed that genetic modification had not produced toxins or allergens that might damage human health. Consequently, there was no need for further tests. As the company told the United Kingdom Advisory Committee on Novel Foods and Process when it applied for a review of the safety of glyphosate tolerant soybeans in 1994, "Following the principles for the application of substantial equivalence, there should be no further safety or nutritional concerns" (quoted in Anderson, 1999, p. 16). Critics disagreed, and in 1995 a coalition led by the major environmental pressure group, Greenpeace, including representatives from consumer and farmers' groups, came together to persuade the country's six major supermarket chains not to stock genetically modified foods until the government enforced labeling and adopted clear regulations on the environmental impact of modified crop plantings (Purdue, 2000). This early initiative established the major axis for the subsequent struggle. On the one side stood the major non-governmental organizations (NGOs) representing environmental, consumer, and farming interests increasingly acting in concert through coalitions built around particular objectives and demands. On the other stood the major biotechnology companies supported by successive governments. Both sides recognized from the outset that commanding public language and imagery was crucial to mobilizing popular support. Consequently, the symbolic politics generated by the struggle between contending frameworks of meaning assumed a central role in the contest.

The fact that unsegregated mixtures of unmodified and modified soy beans were in the process of being imported into Europe from North America became common knowledge in the autumn of 1996, raising concerns that "it was simply a matter of time before the vast majority of the public would find themselves eating an indeterminately large number of unlabeled GM food products, whether they liked it or not" (Durant & Lindsey, 2000, p. 19). However, this did not prevent the outgoing conservative government from raising the permitted levels of glyphosate in soy beans by 200 times in April 1997 (Monbiot, 2001a). A few weeks later, the Conservative Party, traditionally the party most sympathetic to business interests, lost power, swept from office by the landslide victory accorded to New Labor, led by Tony Blair. The biotechnology companies, however, had every reason to believe that despite the change of government it would be business more or less as usual.

Like an earlier Labor Prime Minister, Harold Wilson, who had similarly come to power (in the mid-1960s) after a long period of conservative government, Blair presented himself as a thoroughgoing modernizer promoting a collective future forged in what Wilson had dubbed "the white heat of the technological revolution." The information and biotechnology industries, the two emerging sectors in which British-based research and enterprise enjoyed competitive advantages, were central to the new government's vision of regeneration. Facilitating their expansion was an integral part of their economic strategy. As the publicity issued by the Department of Trade and Industry's Invest in Britain Bureau informed potential overseas investors, "[T]he country ... leads the way in Europe in ensuring that regulation and other measures affecting the development of biotechnology take full account of the concerns of business" (quoted in Monbiot, 2001a, p. 277). Tony Blair had demonstrated his personal commitment to the rapid expansion of the GM food industry at a relatively early point in his first administration. On May 18, 1998, the European Union was due to discuss new measures for labeling genetically modified foods. Hours before deliberations began, Tony Blair had a meeting with Bill Clinton at which, according to the briefing papers, the president pointed out that "the EU's slow and non-transparent approval process for genetically modified organisms had cost US exporters hundred of millions in lost sales" and urged him to "take immediate action to ensure that these products receive a timely review" (quoted in Monbiot, 2001a, pp. 242–243). Britain's representatives responded, using the country's presidency of the EU to counter calls from Denmark, Sweden, and Ireland that all foods derived from engineered DNA or proteins should be labeled as containing genetically modified ingredients (even if those elements were no longer identifiable) and to secure a ruling that only foods in which genetically engineered elements had survived intact should be labeled. As a consequence, consumers had no way to distinguish between foods that might contain high levels of glyphosate used in Roundup Ready soy production and those that were largely free of it.

Pressure groups argued that far from being the triumph for consumer rights to better information announced by the British Agriculture Minister, Jack Cunningham, this rejection of inclusive labeling actively prevented fully informed consumer choice. Their position had already received practical endorsement when Michael Walker, the Chairman of the Iceland frozen food retail chain, announced in April that his stores were banning all GM ingredients from their own-label brands. By tapping into the gathering groundswell of consumer unease around GM foods, he cleverly consolidated the chain's environmentally friendly identity. Yet it was not simply a market ploy. Walker was an active supporter of environmental campaigns and claimed to be the author of the phrase "Frankenstein Foods," which was to play a central role in the subsequent debate. He was, however, an exception. The other major food retailers dismissed comprehensive labeling as unworkable and sought to persuade shoppers that it was unnecessary. The customer information leaflet published by one of the biggest supermarket chains, Tesco, in 1986 was typical. Entitled "Genetically Modified Soya—THE FACTS," it claimed there was no distinguishable difference in the composition of "processed soya beans grown on genetically modified plants [and] those made using conventional soya beans" (quoted in ESRC, 1999, p. 9). This was incorrect. Tests developed by the Ministry of Agriculture, Food, and Fisheries had detected a distinction in compounds in which only 1% of the beans had been genetically modified.

It was against this background of growing mobilization on both sides of the battle for popular trust and support that, in June 1998, Monsanto launched a £1 million advertising campaign designed to "encourage a positive understanding of biotechnology" among the British public. The company's claims were strongly contested by environmental groups. They particularly objected to Monsanto's claim that over its 20-year development of biotechnology techniques it had conducted rigorous tests to ensure its food crops were as safe as the standard alternative. In a ruling published in March 1999, the regulator, the Advertising Standards Authority (ASA), upheld the complaint, pointing out that the company had signally failed to present evidence that it had conducted the safety tests it claimed (Hall, 1999). By the time the ASA reported, however, the tide was already running strongly against both Monsanto and the government.

The potential scale of the company's reversal of fortune had already been outlined by Stanley Greenberg, polling advisor to both Bill Clinton and Tony

Blair, in an internal report to Monsanto presented toward the end of 1998. Surveying the battlefield, he concluded that the advertising campaign "was for the most part, overwhelmed by the society-wide collapse of support for genetic engineering in foods," and there were "large forces at work that are making public acceptance problematic" (cited in Anderson, 1999, p. 115). These forces continued to be resolutely opposed by political supporters of GM foods. In January 1999, the House of Lords European Communities Committee issued a report arguing that the benefits from GM foods greatly outweighed the risks, and the government launched its "Biowise" initiative, committing £13 million of public investment to Britain's biotech industry over 4 years. These positive moves made little impact. Over the next few weeks, opposition became ever more visible and vocal, with the major daily newspapers taking a leading role in articulating public disquiet. Debate centered on two main issues: food safety and environmental impacts.

OF RATS AND MEN

> For many Europeans, the first encounter with GM technology was being told they were already eating GMOs. It was literally and figuratively being "shoved down their throats" by the companies with little or no regulation or public consultation. . . . If I'd been told the food I was eating recently was treated in a new, potentially hazardous manner, I'd vigorously . . . demand a time-out to investigate the matter too. (Allan McHughen, Canadian Research Scientist; cited in McHughen 2000, p. 111)

Questions about the possible health risks of genetically modified foods had first reached a wider public in Britain when a research scientist, Arpad Pusztai, appeared on one of the country's leading TV current affairs programs, *World in Action*, in August 1998. Pusztai expressed concern that recent laboratory experiments he had conducted with rats suggested that feeding them with genetically modified potatoes prompted changes in vital organs and a weakening of the immune system.

Pusztai was a respected scientist, a distinguished researcher, and an acknowledged expert on lectins—proteins that can help plants resist attacks by insects. He was a long-standing member of the Rowett Research Institute whose history illustrates well the century-long development of biotechnology research outlined by Leah Lievrouw in her chapter. Established in 1913 by the wine grower, John Quiller, it emerged to become a major public research institute. In 1995, Pusztai's team successfully bid for a major research contract from the Scottish Office. They set out to investigate whether potatoes, a staple of the British diet, could be modified to resist insects while still being safe to eat. They generated relevant data by feeding differently treated potatoes to groups of rats. One group was given undoctored potatoes, another potatoes that had been treated with the snowdrop lectin, and a third potatoes that had been genetically engineered to produce the snowdrop lectin. When he began the study, Pusztai was concerned to discover that he could locate only one refereed journal article exploring the potential risks of genetically modified foods and this had been produced by a researcher employed by Monsanto. Even so he saw no reason to alter his general opinion that genetic modification was a generally positive intervention. As he later told a journalist, "I was absolutely confident I wouldn't find anything. But the longer I spent on the experiments, the more uneasy I became" (quoted in Lean, 1999, p. 7). By this time, however, the money for the study had run out with little prospect of a follow-up grant. Convinced of the need for further research, Pusztai set about trying to raise funding from other sources. Appearing on a widely watched prime-time TV program offered a chance to catch potential sponsors' attention. As he later put it, "We were not getting any more money. . . . This was a way of crying for help" (quoted in Hindle, 1999, p. 11). His appearance had the Institute's full backing, but in making the case for further research he used a form of words that cut directly across government and food industry attempts to reassure the public, arguing that he found it "very unfair to use our fellow citizens as guinea pigs" and that "we have to find [experimental subjects] in the laboratory." As he later told a Parliamentary Committee on Science and Technology, "I thought it was a fair comment, not a wise comment, but a fair one" (quoted in Radford, 1999, p. 15). However, it was sufficient to precipitate his suspension from the Institute, deny him further access to his data, and prompt an internal inquiry into the validity of his procedures and results, which duly concluded that he had failed to establish a link between genetically modified potatoes and potential risks to human health.

Required not to speak publicly about the research, Pusztai was unable to respond to this report when it appeared 11 days after his TV interview, and the story disappeared from public view. It was reactivated early in February 1999 when 22 international scientists issued a public statement challenging the conclusions of the internal audit report, deploring the way Pusztai had been treated and arguing strongly for further research. As one signatory to the statement, Professor Ian Pryme of Bergen University, put it, "There can be little doubt that in light of the available data further detailed experimentation is certainly warranted" (quoted in *The Guardian*, 1999, p. 7). Faced with this unexpected endorsement of Pusztai's questioning of official reassurances on the safety of GM foods, the scientific establishment mobilized to undermine his credibility. The Royal Society, Britain's most prestigious scientific organization, appointed a committee of six scientists to evaluate his findings. They reported in May, arguing that they found his work "flawed in many aspects of design, execution and analysis and that no conclusions should be drawn from it" (quoted in Connor, 1999a, p. 10). Because the composition of the committee's members was not made public, it was impossible for critics to judge their impartiality, although the close links that key Society officials enjoyed with both the biotechnology industry and relevant government departments raised doubts. At the time, for example, the Society's science policy division, which oversees its biotechnology policy, was headed by Rebecca Bowden, who had moved for the government's biotechnology unit at the Department of the Environment. The Royal Society returned to the fray in October when its president, Sir Aaron Klug, publicly attacked the country's leading medical journal, *The Lancet*, for printing a paper by Pusztai and according his work "an authenticity it doesn't deserve" (McKie, 1999, p. 10). These exceptional interventions underscore the growing politicization of scientific inquiry in an era in which commercial interests are shaping research agendas and publics are increasingly demanding the minimization of risks—a tension we return to presently.

In retrospect, the Royal Society's interventions appear as a rearguard action. Pusztai's concern that members of the public were being used as guinea pigs without their knowledge had already become common currency. Four days after the international statement supporting him, the country's second largest circulation daily tabloid newspaper, *The Daily Mirror*, launched a campaign for comprehensive food labeling entitled "Label Frankenstein Foods" under the slogan "Tell us all the facts." They joined another mass circulation popular daily, *The Daily Mail*, which had introduced their "Gene Watch Campaign" some days earlier with a story reporting their own laboratory tests headed, "Top brand foods made with mutant crops, and the companies don't even know it." The opening line claimed that, "The creeping progress of 'Frankenstein' ingredients into our diets is exposed again today with laboratory tests" (Poulter, 1999, p. 7).

The fact that newspapers traditionally located to both the left of the political center (*The Mirror*) and the right (*The Mail*) should organize their arguments around the figure Frankenstein is no accident. Mary Shelley's cautionary tale of a scientist who sets out to discover the secret of life and succeeds only in creating monsters has become "the governing myth of modern biology" and one of the most ubiquitous images in popular culture (Turney, 1998, p. 3). It integrates deep-seated fears about crossing natural boundaries with generalized concerns about the unlooked-for risks of new processes and expresses them in an immediately recognizable image that crosses conventional political divisions and addresses readers as consumers with common interests in personal safety and quality of life. To dismiss this coverage as sensationalist is to fundamentally misunderstand how popular communication works. Constructing accounts that achieve widespread currency requires the mobilization of resonant images as well as the transfer of information and orchestration of argument. Symbolic work is not a distortion or an addition to the process of mediated sense making. It is integral to it.

At the same time, producing news also involves telling stories that make sense of novel events by presenting them as the latest episodes in an unfolding narrative organized around stock themes and characters. In the case of the GM debate, the saga of Britain's experience with bovine spongiform encephalopathy (BSE) in cattle (or mad cow disease as it was popularly known) provided a particularly potent narrative context. The period over which the official inquiry into the BSE outbreak sat coincided almost exactly with the initial debate over Pusztai's findings, and the evidence it gathered offered tempting parallels between the two situations. Both involved the transgression of natural boundaries by turning grazing animals into meat eaters in the one case and incorporating genes from one species into another on the other. Both told of early warnings of risk being ignored, efforts to discredit critical scientists, and government inaction and cover-up. Unlike the GM food story, the BSE story had an ending—the eventual emergence of a link between BSE in cattle and a fatal wasting condition, variant Creutzfeld-Jackob disease (vCJD), in humans transmitted through the food chain. As the national broadsheet daily, *The Independent* (1999), argued in an editorial headed "The Merits of Letting us Know What We Are Eating," published 4 days after the international statement supporting Pusztai,

> [T]his is a field of science in which, as the BSE crisis revealed, the speed of discovery is matched only by the speed with which the widening vistas of human ignorance are opened up. Mad cow disease turned out to be transmittable by means previously thought impossible, and is still not fully understood. (p. 3)

The salience of BSE in framing concerns about GM foods in the first, most intensive, phase of the GM debate of 1999 is confirmed by research showing that it was mentioned in as many as 25% of the items that appeared in major national newspapers (Durant & Lindsey, 2000, p. 11). As with other aspects of the debate, the comparison with BSE was anchored by imagery as well as language.

In February, when the debate was at its most intense, the prime minister reaffirmed his strong support by claiming that his own family was happy to eat GM foods. On February 15th, *The Independent* carried a large cartoon captioned "The PM says GM's OK," showing him and his wife standing in Downing Street in front of a battery of news crews ramming a potato into the mouth of one of their startled children whose thought bubble says, "I wanted a beefburger." British readers would immediately recognize this as a lampoon of the best-known news photo from the BSE crisis showing the then agriculture minister, John Gummer, feeding his young daughter with a beefburger at a country fair at a time when concerns over meat safety were growing. The implied equivalence was clear: "They lied to you then, they are lying to you now." Not surprisingly, the government objected strongly to these kinds of representations, and in April, Jack Cunningham, the Cabinet Office Minister then responsible for coordinating official GM policy, accused the media of engendering "mass hysteria" and "warned that the GM food issue should not be compared with the BSE catastrophe" (Durant & Lindsey, 2000, p. 10). Again this is a misunderstanding of the dynamics of popular meaning making.

Given the living legacy of BSE portrayed in the harrowing news pictures of young people dying a slow and painful death from vCJD, we can argue that far from being an irrational media-driven overreaction, popular opposition to GM foods was based on an entirely plausible extrapolation from past experience and the deepening distrust of government intentions it had engendered. As one of Britain's best-known journalists noted, people had spent "a lifetime listening to experts tell us there is no risk—only to discover years later that many of them were wrong all along" (Humphrys, 2002, p. 201). When a general sample of the adult population was asked in January 1999 which sources they would trust to advise them on the risks posed by BSE, only 4% nominated government ministers and 17% government scientists. In contrast, 57% nominated independent scientists working in universities or independent research institutes (House of Lords, 2000, Appendix 6). Supermarkets and food manufacturers also fared badly, with only 6% and 11% of respondents, respectively, prepared to trust them. Given the striking similarities between the issue careers of BSE and GM foods, this established bedrock of skepticism was easily carried over to concerns about the possible health risks of GM foods and readily converted into support for campaigns for comprehensive labeling.

Faced with mounting consumer discontent and boycotts, almost all the major supermarket chains had followed Iceland in banning GM ingredients from their own-brand products by the end of March. They were joined the next month by a succession of leading food manufacturers, including Unilever, Nestle, and Cadbury. By June even the staunchest supporters of GM within the food industry had reluctantly accepted the commercial wisdom of banning GM ingredients from their products. As the chief executive of Northern Foods, one of the technology's most vocal advocates, put it, "We support GM foods in terms of safety. But we have to take our lead from our customers, and our customers want non-GM foods" (quoted in William, 1999, p. 11).

The struggle for the supermarket shelves was only one front in the battle over GM foods, however. A second front had also been opened around the risks that GM plantings posed to the environment and alternative forms of agriculture, particularly organic farming. This battle was to be even more bitterly fought, with protest tactics ranging from concerted lobbying to direct action to destroy trial crops.

KILLING FIELDS

There have been 25,000 trials of GM crops in the world now, and not a single incident, or anything dangerous in these releases. You would have thought that if it was a dangerous technology there would have been a slip up by now. (Thomas Joliffe, Research and Development Manager for Adventa Holdings UK; quoted in Anderson, 1999, p. 31)

Led by analysts in the United States, where there were extensive commercial plantings of GM crops, concerned scientists identified several potential environmental hazards (Rissler & Mellon, 1996). First, there is the possibility that birds and other wildlife will be adversely affected by the disappearance of their traditional habitats and by ingesting the active elements in GM crops—a risk that resonates strongly with long-standing concerns over the growing list of endangered species and the progressive reduction of ecological diversity already caused by the rapid growth of intensive and single-crop farming practices. This interpretive frame was introduced early on in the debate and articulated particularly strongly by national dailies with a strong commitment to ecological issues, with headlines such as "Gene crops could spell extinction for birds" (Carrol, 1999, p. 9) and "Modified pollen kills threatened butterfly" (Connor, 1999, p. 1). This second item summarized an article being carried by the respected scholarly journal *Nature*, reporting a study conducted by John Losey and his colleagues at Cornell University. The quoted comment from one of his coworkers, Dr. Linda Rowe, encapsulated the objectors' central concern: "Monarchs are considered to be a flagship species for conservation. This is a warning bell" (quoted in Connor, 1999, p. 1).

These concerns were strengthened by the possibility that traits from GM crops might be transferred to other plant varieties sharing the same habitat, creating new species of unwanted superweeds that would squeeze out the original inhabitants, further reducing biodiversity. This was not simply an environmental concern—it had real commercial consequences. If gene drift spread to fields planted with organic produce, British growers would be barred from claiming organic status and excluded from a lucrative and rapidly growing segment of the food retailing market.

As the quotation that heads this section makes clear, industry advocates dismissed these concerns, arguing that evidence from trials conducted elsewhere showed they were unfounded. British critics, however, disagreed, maintaining that environmental impact evaluations undertaken in the United States had never been designed to assess the full range of possible effects, and that differences in the organization of farming between the two nations called the relevance of American evidence into question. Unlike the large and relative isolated farms of the U.S. Midwest, for example, British farms tend to be smaller, more crowded together, and more likely to have boundaries with woodlands and other major wildlife habitats.

Environmental concern first reached the public domain in June 1998, when Prince Charles, an enthusiastic practitioner of organic farming on his own estate at Highgrove, called for a public debate on the merits of allowing GM crops to be grown in Britain. The following month, the government's official advisory agency on wildlife conservation, English Nature, called for a 3-year moratorium on commercial planting. These calls resonated with deep-seated cultural motifs and values. Idealized visions of the rural landscape play a central role in the British sense of national uniqueness and are expressed in multiple popular forms, from mass reproductions of John Constable's best-known painting "The Hay Wain," to the illustrations in children's story books, to the well-worn phrase "green and pleasant land" taken from William Blake's poem "Jerusalem," often sung at public gatherings as a second national anthem.

In February 1999, English Nature's Chair, Baroness Young, repeated the organization's demand in an article published in the national broadsheet, *The Independent on Sunday*, arguing that the 1-year voluntary period agreed on between the government and biotech industry was totally inadequate for a proper assessment. The following week, the Conservative Party, sensing an issue that might boost their low public opinion poll ratings, announced that they would introduce a Parliamentary Bill calling for a 3-year ban. Once again, however, campaigning groups seized the initiative by forming a broadly based coalition to lobby for an extended moratorium. This 5-Year Freeze campaign mobilized not only environmental and nature conservation groups, but a wide range of traditionally moderate citizens' organizations such as the Federation of Women's Institutes and the Townswomen's Guild. By constructing this broad base of representation, the NGOs successfully outflanked the government, making it increasingly difficult for them to argue that the call for a moratorium came simply from special interest groups. Their case was bolstered when the result of a court case brought against Monsanto by the government oversight body, the Health and Safety Executive, was published on February 17, fining the company £17,000 for "partially removing" the control measures designed to prevent a trial GM planting from cross-pollinating neighboring plants. Faced with this public evidence of corporate noncompliance, the government conceded, and on February 18, the environment minister, Michael Meacher, announced that, "Until we have clear scientific evidence about the impact on the environment we will continue to prevent the commercial planting of these crops as long as possible" (quoted in Brown, 1999, p. 1). In practice, however, this open-ended commitment was translated into a 4-year period of intensive trials.

Critics responded by arguing that the relatively limited nature of the tests the government envisaged conducting would not produce sufficiently clear evidence to settle the issue. This claim received authoritative backing in June from a study conducted for the government by the John Innes Centre, the country's leading research institute in the field, which concluded that, "easy and reliable methods of identifying and quantifying GM contamination ... may be very difficult to achieve" (quoted in Meikel, 1999, p. 6). The Environment Minister again adopted a conciliatory tone, announcing that, "We should not be stampeded by industrial or commercial interests to take a decision in favour of this type of technology until we know and produce evidence for everyone to look at that it is wholly safe" (quoted in Arthur, 1999, p. 8). By this time, however, the debate had already moved on, with those opposed to any introduction of GM crops resorting to direct action, uprooting and destroying trial crops across the country.

The activists had gained a significant public relations victory in March when the prosecution brought against two women who had destroyed a trial crop of genetically modified maize in the West Country was dropped on the advice of the Director of Public Prosecutions, following the defense submission of 10 expert reports detailing the risks posed by the technology and the failures of the regulatory system. As the women's solicitor [attorney] pointed out, "By withdrawing the case ... the Crown have accepted that there was compelling evidence that the defendants had a lawful excuse to remove the GM maize" (quoted in Anderson, 1999, p. 116). In another well-publicized trial the following month, Monsanto suffered a severe setback in the High Court in pursuit of an action it had brought against activists from the group GenetiX Snowball. The judge refused the company's request for a permanent injunction preventing the defendants from damaging crops in the future and ordered that the case proceed to a full trial on the grounds that the defendants had an arguable case that they were acting in the public interest. This would allow the accused to call expert witnesses, effectively putting the company on trial.

Crop destruction continued over the following months. Some incidents merited only local reports, but others attracted national publicity. The most extensively covered occurred at the end of July, when Lord Melchett, a landowner, former minister in the Labor Government of Harold Wilson, and executive director of the leading environmental pressure group Greenpeace, led a group of protesters dressed in the kind of decontamination suits worn at crime scenes and nuclear accidents into a field near his own estates in Norfolk and uprooted a trial planting of genetically modified maize. This carefully crafted theatricality was typical of the group's mastery of the photo opportunity. In another widely publicized incident in February, at the peak of the first phase of debate over GM foods, they dumped four tons of genetically modified soya beans outside the Prime Minister's official London residence in Downing Street. The truck bore the slogan: "Tony, Don't Swallow Bill's Seed." The high visibility of the location and the humorous reference to the Monica Lewinsky scandal guaranteed that it would be widely covered by press photographers and TV news teams. The field invasion, in contrast, took place in a relatively isolated rural area. Consequently, to ensure it was covered, the organizers produced their own photographs and videotaped record of the action, which they then offered to news organizations.

As a result of the action, Lord Melchett was arrested and briefly held in custody before being released. In an article published in *The Independent* 2 days later, he presented direct action as the only way to redress the "democratic deficit" created by the government's failure to defend the public interest:

> [N]on-violent direct action . . . makes clear when a democracy is failing. Astonishingly, the peaceful removal of GM crops before they flower is practically the only democratic veto UK citizens currently have to prevent genetic pollution . . . At no point have the people given their consent. . . . The private interests of a small handful of chemical corporations have been raised above the public's right to an uncontaminated environment and access to non-GM food. (Melchett, 1999, p. 24)

His reading of the public mood received public support when the case came to trial the following year; after a lengthy process costing the Crown Prosecution Service an estimated quarter of a million pounds, the jury found the defendants not guilty on the principal charge of criminal damage. As one of the solicitors representing the protesters noted, "Juries understand reasonable citizens actions" (quoted in Kelso, 2000, p. 1).

Well before this, however, it was becoming increasingly evident that concerted opposition to GM foods from an unprecedented array of pressure groups, coupled with the widespread public distrust demonstrated by successive opinion polls, was placing major barriers in the way of the agrichemical companies' expansion plans and seriously damaging their public image. As the best-known corporation operating in the field, Monsanto was particularly affected, seeing the value of its stock fall by 11% in the 6 months prior to September 1999. The company's chief executive officer, Bob Shapiro, recognized that they had badly mismanaged the situation, admitting to a business conference in London in October that, "We have irritated and antagonized more people than we have persuaded," and that, "Our con-

fidence in biotechnology has been widely seen as arrogance and conde-
scension" (quoted in Vidal, 1999b, p. 11). Yet the damage was already done.
In December, the contract caterers serving the company's main offices in
Britain decided to ban GM foods from the staff canteen in response, as they
put it, "to concerns raised by our customers." The following month, Monsanto merged with the U.S.–Swiss drugs company, Pharmacia and Upjohn.
The new enterprise was named Pharmacia, and the Monsanto brand was
relegated to an agricultural subsidiary.

In the space available here, I have only been able to sketch the general
outlines of the great GM foods debate that took place in Britain in 1999. Nevertheless, I hope I have said enough to convince you that it offers a particularly rich case study of the emerging politics around biotechnologies and
raises multiple questions about the shifting relations among science, government, communication circuits, and the public. I want to focus on two:
the changing nature of scientific inquiry and debate under conditions of increasing commercialization and media saturation; and the role of communication and information media—particularly popular imagery—in constructing the symbolic field on which contending interests compete for public
attention and trust.

POSTNORMAL SCIENCE AND INFORMATION CAPITALISM

Together with entrepreneurs, scientists were the heroes of modernity's master narrative of progress. Their discoveries promised to release people from
the fear of hunger and disease and deliver increasing levels of comfort and
choice. This tale of incremental improvement did not go unchallenged, however. It was continually dogged by cautionary tales of creations turned
against their makers. Many were variations on the story of the Golem, an ani-
mated clay figure built to do menial tasks. In the earlier medieval versions of
this originally Jewish tale, these figures appear as benign forces relieving
drudgery and protecting the community against persecution. Yet by the time
Johann Schmidt published his survey of the myth in 1682, he found they were
rebelling against their servitude and "inflicting great damage upon the person of their master" (Jewish Gothic, 2001, p. 1). It was this unruly figure that
Mary Shelley took as her model for the creature in *Frankenstein* and that became in turn the central metaphor for science's inability to cope with the unanticipated consequences of its interventions. This heightened sense of risk
has been powerfully reinforced in recent years. After the malfunctions at the
Three Mile Island and Chernobyl plants, nuclear power, which had been vigorously promoted as a cheap, clean, and safe source of energy, appeared as
more of a problem than a solution. The automobile and refrigerator, which

had been sold as unalloyed improvements to the quality of life, were found to be contributing to a widening hole in the ozone layer and growing climatic instabilities. However, it was the two technologies most closely associated with the emerging order of information capitalism and most comprehensively integrated into everyday life—computers and genetically modified foods—that proved the most potent focus for anxieties.

When the first widely available, genetically modified (GM) food product, the Flavr Savr tomato, was introduced in 1994, it was met with almost universal acclaim and was welcomed even by newspapers like *The Guardian*, which later adopted a much more critical stance toward GM technologies. It was a brilliant choice as the "vanguard" of the GM revolution because "tomatoes are everywhere—in salads, . . . in sauces and on pizzas, in pies and sandwiches" (Myerson, 2001, pp. 11–12). Yet this ubiquity also pointed up the generalized nature of possible risks suggesting that consumers would be unable to avoid them even if they so desired. By 1999, the tomato had become a widely adopted sign of uncertainty, providing the graphic anchor for three of the national press campaigns launched in relation to GM foods. They ranged from *The Independent's* simple drawing of a ripe tomato, to the *Daily Express'* picture of a tomato being examined under a microscope, to the *Daily Mail's* logo showing a tomato with a jigsaw-shaped piece removed—a particularly potent image of unresolved questions.

The increased questioning of GM technologies coincided with deepening anxieties over possible computer malfunctions. A succession of viruses that had destroyed data and damaged software on desktop machines had already demonstrated the network society's vulnerability, but it was not until fears over the possible impact of the "Millennium Bug" began to gather momentum in 1999 that concern became generalized. In the same way that the argument over genetically modified soya had shown how thoroughly GM ingredients were already integrated into everyday food products, so the mounting media coverage of the "Bug" made clear how ubiquitous computer technology had become and how a malfunction would impact everyone regardless of whether they owned a computer. As a British government leaflet, issued to businesses in mid-1999, explained, "There are potential problems in every type of programmable electronic system—computers, microprocessors, and the embedded systems ('chips') built into many modern instruments, controllers and machinery . . . lifts and escalators, fire detection and alarm systems" (Health and Safety Executive, 1999, p. 3). Here was a classic case of unanticipated consequences and foreshortened time scales. A routine design decision to limit the number of digits computer clocks could register, taken when computer applications were still relatively limited, seemed capable of paralyzing the entire social system, cutting off power, scrambling financial transactions, and knocking out a range of household appliances. In the event (and after the devotion of enormous resources to prevent the problem from having any effect), the predictions proved unfounded. As the British Government Millennium Centre (GMC) reported on January 1, 2000, "All appears quiet both in the UK and abroad. There have been no manifestations of bug-related problems" (Government Millennium Centre, 2000, p. 1). The relief here is audible, but the fact that the government had set up a special unit within the Cabinet Office to "provide Ministers, the media and the public with details of what is happening and how any problems are being dealt with" (Government Millennium Centre, 1999, p. 1) demonstrates how seriously they took the situation and how unpredictable the impacts remained right up until the strike of midnight.

The BBC was certainly prepared for the worst. The back wall of its millennium night studio was dominated by a powerful image of the expected tidal wave of chaos in the form of a huge photograph of a sinister-looking insect and a line moving across a map of the world as countries reached midnight. Live coverage of festivities around the globe was regularly interrupted for updates on the global "Bug" situation and the locations of reported computer malfunctions displayed on the map.

As noted earlier, the growing recognition that all interventions are characterized by unavoidable uncertainty and unpredictable impacts has undermined both the unchallenged authority of scientists and the assumption that innovations always accelerate social progress. In this changed situation, both scientists and governments are faced with a much more skeptical public. As the British Government's Chief Scientific Advisor, Professor David King, conceded, "We have seen a transition from a time when a scientist would have been respected automatically. Now people are questioning that authority, and demanding to be consulted about major innovations" (quoted in Plomin, 2001, p. 31). Many observers welcome this trend and look forward to a greater "democratization of science ... bringing science into public debate along with all the other issues affecting our society" (Funtowicz & Ravetz, 1992, p. 254). In the age of progress, governments were judged on how equitably they distributed benefits. In the era of generalized risk, they come to be judged on their capacity to reduce avoidable threats (see Beck, 1992). Because decisions on risks involve difficult questions of values and interests—of who gains and who loses—they have inevitably moved more and more toward the center of public debate.

One way to address the radical uncertainty surrounding key technological innovations is to adopt the precautionary principle. As the French Conseil d'Etat put it when responding to a scandal involving contaminated blood, "In an uncertain situation, an hypothesis [of possible harm] that cannot be rejected should be taken as temporarily valid, even if it cannot be formally proven" (quoted in Gollier, 2001, p. 309). This position was strongly supported by an international meeting of scientists, lawyers, and environmentalists who argued in a statement issued in January 1988,

... people must proceed more carefully than has been the case in recent history.... When an activity raises threats of harm to human health or the environment, precautionary measures should be taken even if some cause and effect relationships are not fully established scientifically. (quoted in Anderson, 1999, p. 33)

Advocates of this principle argue that it applies with particular force in the area of GM foods because introducing any modification has the potential to generate a chain reaction that ripples through an ecological system in unforeseen ways, causing damage that may not become evident until it is too late to rectify it. As the food writer, Joanna Blythman, argued in an article in *The Independent on Sunday* early in the British debate of 1999, the GM products already on the shelves represent "the initial applications of a new, unpredictable technology which when it goes wrong—and it already has—could have catastrophic, irreversible consequences.... When we start tinkering ... even the cleverest geneticists cannot predict the knock-on effects" (Blythman, 1999, p. 8).

The fact that the precautionary principle places the burden of proof squarely on the shoulders of those proposing innovations cuts sharply across corporate desires to exploit potentially profitable knowledge as quickly and widely as possible. Promoting this "just do it" stance requires companies to reaffirm the links between scientific innovation and social progress. As a press advertisement issued by the pharmaceutical company, Pfizer, which appeared in a number of publications at the height of the 1999 GM foods debate put it, "Experience demonstrates that the risks of innovation, while real, are vastly less than the risks of stagnation.... Allowing the Precautionary Principle to dominate decision making ... could suppress the very forces of economic and technological innovation that make the current world possible." This position was strongly supported by Lord Sainsbury, the Minister for Science and Innovation, who dismissed opponents of GM foods as mindless opponents of progress.

The benefits of science have been huge: we should not deny them to future generations.... I am not arguing for the mindless pursuit of economic and social change; I am arguing against mindless opposition to it.... We cannot turn away from progress, but we can encourage it and allow other countries to enjoy its advantages. (Sainsbury, 1999, p. 28)

For many British commentators, this general argument was further underlined by the logic of national competitive advantage. They saw "the future of biotechnology in Britain, one of [the country's] most important and promising industries," needlessly imperiled by the protesters' refusal to accept that "there is no evidence of danger to health from GM crops"

(Taverne, 2001, p. 58). This formulation places the onus of proof back on the shoulders of critics. If practical experience shows that there is no sign of a technology being broke, supporters argue, why fix it? As Maarten Chrispeels of the University of California–San Diego bluntly put it, "Twenty million hectares of GMO's are being grown in the United States and nobody is turning into a tulip" (quoted in Radford, 1999, p. 17).

The dispute over the precautionary principle goes to the heart of capitalism's ongoing shift from a formation based on industrial production to a system organized around the command of strategic information. As one of the primary building blocks of this emerging information capitalism, biotechnologies are at the center of contemporary struggles between corporate interests and the common good. To maximize their advantages on this new economic playing field, corporations have made strenuous efforts to monopolize the ownership of strategic knowledge and direct the ways it is exploited. In pursuit of this goal, they have employed three main tactics: enclosure, bundling, and co-option.

Enclosure

A number of observers have compared the conversion of vernacular botanical and biological knowledge into patentable discoveries to the English enclosure movements, which fenced off traditional common lands, incorporating them into private estates and prosecuted anyone who trespassed on ground to which they once had enjoyed free access. In the process, what had previously been a shared asset and a common heritage became a commodity to be owned and traded (Rifkin, 1999). In the field of biotechnology, strategic knowledge has been progressively fenced off by new forms of intellectual property rights, which have granted patents to novel applications of the properties of plants and living organisms on the grounds that they are inventions that add value rather than products of nature. As a consequence, increasing stocks of folk knowledge from the biodiverse countries of the South have been appropriated by major companies in the North, who then sell the newly patented products based on this information back to these countries. The influential Indian activist Shiva (1998) saw these moves as simple acts of plunder by buccaneering companies and described them as *biopiracy*–a label that has achieved considerable currency in debates. One way to ensure that users respect newly created intellectual property rights is to incorporate them into the composition of the product. Seeds are a particularly contested case of this.

Traditional agricultural systems are based on saving seeds from the harvest for planting the following year and on pooling local knowledge of which varieties grow best in particular conditions. These practices leave little room for commercial marketing, however. The solution is to sell hybrid seeds produced by crossing two varieties. Only a proportion of the seeds collected from these crops will display the same characteristics as the originals, with the rest displaying less desirable traits. Alternatively, seeds can be rendered sterile so that they cannot produce viable seeds after cropping. In both cases, buyers are compelled to buy a new set of seeds each year because knowledge of the makeup of the parent seeds used to produce the original hybrid is monopolized by the company (Purdue, 2000). The production of sterile or *terminator* seeds (as they have been dubbed) has been fiercely contested. In October 1999, in another setback, Monsanto bowed to international pressure and agreed to abandon its plans to commercialize terminator technology, although it continued to develop other seed control techniques (Shiva, 2000).

Struggles between communality and intellectual property rights are also a major feature of computerized information systems. Corporate-owned computer software packages operate with a proprietary model that keeps the source code, which determine how they operate, secret; this makes it impossible for users to fix bugs, alter programs to fit their own requirements, or distribute the code to other users. These efforts to exercise total control over software design and use, however, had been consistently opposed by the development of public domain software which is both freely available and collectively modified and improved in practice by the user community. The battle over UNIX, one of the major operating systems, has been particularly fierce.

UNIX was developed in the Bell Laboratories run by AT&T, the regulated monopoly that operated the U.S. telephone system. When the corporation was divested of some of its core telephone activities in 1984, it needed to find other profit centers and was granted proprietary rights over UNIX. In response, Richard Stallman, a programmer at MIT, launched the Free Software Movement based on the principle of *copyleft* as opposed to *copyright*. This required anyone using software that had been made freely available to distribute any improvements they had made to the source code over the internet. The most successful attempt to do this was initiated by a graduate student at the University of Helsinki, Linus Torvalds. His operating system, Linux, has been constantly upgraded by the work of thousands of hackers and millions of users to the point where it "is now considered one of the most advanced operating systems in the world, particularly for internet-based computing" (Castells, 2001, p. 14). Nor is it simply the preserve of small concerns and recent entrants to the market. Major companies, including IBM, have increasingly migrated to Linux as an operating system, prompting Microsoft's CEO to complain that Linux has become "a cancer that attaches itself in an intellectual property sense to everything it touches" (quoted in Moody, 2002, p. 3).

These instances point to the difficulty of defending property rights under information capitalism. Knowledge systems are much more leaky than factories and ideas more mobile than physical goods, which is why companies look for other ways to lock consumers into their products. (This problem is explored in greater depth by May, chap. 5, this volume.)

Bundling

As noted earlier, Monsanto and a number of other major companies involved in developing GM foods are primarily pharmaceutical and chemical producers; they have engineered crops like Roundup Ready soy beans, which are designed to resist their own proprietary herbicides and weed control systems. This produces two income streams: sales of modified seeds and sales of the herbicide that will eradicate weeds in planted fields while leaving the crop untouched. By rendering other methods of weed control less effective and more costly, bundling products together in this way massively increases buyers' dependence on the company and locks them into agricultural practices determined by boardroom calculations of profit rather than farmers' assessments of the cultivation methods best suited to local conditions or requirements. As critics have argued, the inordinate control of the food chain that GM crops potentially grant to a handful of companies has "grave implications for food security" (see Monbiot, 2001b, p. 30), particularly in low-income countries that previously relied on subsistence agriculture, but are now being urged to produce cash crops for the world market.

Again we can see this same abuse of market power at work in the information industries. The U.S. government's antitrust action against Microsoft, for example, centered on claims that the company had used the dominant position of its Windows software (which is loaded on over 90% of all personal computers around the world) as a platform to unfairly promote its other products, particularly its internet browser, Explorer. By bundling these additional programs in with the basic Windows package, displaying their icons on the welcome page, which appears on the screen automatically whenever the machine starts up, and excluding rival products such as the Netscape browser, Microsoft, like Monsanto, attempted to lock users into its own range of products and lock out the competition.

Cooption

Enclosing access to strategic knowledge through patents and property rights and maximizing the profits made on its applications through bundling have helped considerably in consolidating corporate control over the uses of both information technologies and biotechnologies. However, corporations are not the only agencies involved in the generation of new knowledge. Universities and publicly funded research institutes remain important centers of novel ideas that must also be enlisted into the service of profit maximization. This process of cooption has been accelerated by persistent cuts in universities' core budgets over the last 2 decades and mounting pressure on researchers to compensate for shortfalls in public subsidy by earning more income from commercial activities. As a consequence, the scientific community has become enmeshed in an ever-widening range of public-private partnerships. These can cover almost all aspects of scientific work. As the former editor of the *New England Journal of Medicine*, Marcia Angell, noted, "Researchers serve as consultants to companies whose products they are studying, join advisory boards and speakers' bureaus, enter into patent and royalty arrangements, agree to be listed authors of articles ghost-written by interested companies" and may "also have equity interest in the company" (quoted in Boseley, 2002, p. 4).

Given the centrality of experiments on mice and rats in the life sciences, it is perhaps no surprise that OncoMouse, the laboratory mouse redesigned to produce cancers that can be investigated in search of cures, has become a particularly iconic example. With a patent held by Harvard University and commercial rights held by DuPont, this genetically engineered rodent is simultaneously "an academic product, a commercial commodity, a research tool and a space in a sales catalogue" (Myerson, 2001, p. 51). As Lievrouw points out (chap. 6, this volume), the splicing together of public science and corporate strategies has had important consequences for scholarly communication and exchange. Knowledge with commercial potential is increasingly restricted to a designated list of trusted insiders rather than being openly published. It circulates within a concealed club of interested parties rather than an invisible college of fellow scholars. As noted earlier, when Arpad Pusztai was planning his research on the possible health risks of eating GM foods, he could only find one relevant refereed scientific article on the topic. Monsanto may, as they claimed, have conducted extensive research on the possible health risks of GM foods, but the results were not publicly available.

Despite concerted efforts, however, the corporate cooption of scientific inquiry and debate has not been as comprehensive as some commentators have suggested. One reason for this is the contemporaneous growth of the internet, which offers an alternative network for distribution and discussion. As soon as Pusztai regained access to his contested research notes, for example, he published them on his own website.

Supporters of introducing the commercial planting of GM crops in Britain as rapidly as possible have often labeled their critics as *antiscience*. Opinion polls, however, consistently show that people draw a sharp distinction between independent scientists working in universities and those working for corporations or governments (see e.g., House of Lords, 2000). They believe university researchers are doing their best to establish the truth in the public interest, but they suspect that the other two groups speak with their master's voice. The issue then is not whether people are "for" or "against" science in the abstract, but who they trust to tell the truth about its likely risks and costs.

This increased contestability of claims is partly a product of the escalating uncertainties surrounding scientific conjectures in postnormal times, but in the field of biotechnologies it is reinforced by the growing politicization of consumption, which continually pitches the marketing industry's promises of enhanced health and well-being against the uncomfortable reminders of potential harm and damage publicized by environmental and consumer activists. In the ensuing struggle for credibility and public support, command over the way issues are framed and anchored in resonant images plays a pivotal role.

ARGUING WITH IMAGES: FRAMES AND PICTURES

Most accounts of the mass media's role in framing accounts of risk have tended to focus on the way language is deployed to bolster or contest particular positions and have paid little or no attention to visual imagery. This linguistic domination of research is partly due to the rapidly expanding range of analytical tools provided by critical linguistics, discourse studies, and conversation analysis, which offer well-oiled machines for producing publishable papers. Yet it is also written into ways that dominant models encourage us to think about communication processes. Transportation models, which see communication as a matter of moving information from one point to another, think of meaning as packaged in *messages*—a metaphor that inevitably pulls attention toward written materials and leaves little room for sustained analysis of the visual elements in news.

The influential social amplification of the risk model developed by Roger Kasperson and his colleagues is a case in point. They acknowledged that the "symbolic connotations" attached to widely reproduced images can play a major role in stigmatizing particular technologies (Kasperson et al., 1988). They are particularly concerned that the images of mushroom clouds and workers with gas masks habitually featured in popular fictions and films about nuclear power act as a subliminal "message transfer" system connecting the technology to "frightening symbols" (Renn, 1991, p. 298), making it more and more difficult for the industry to convince the public that it can provide safe energy. However, because their conceptual framework is based on an information transfer model derived from engineering they are led to focus on the manifest content of media messages (see

Murdock, Petts, & Horlick-Jones, 2003). As a result, the latent, connotative meanings carried by cultural symbols appear as an optional add-on or extra—something that "may be attached to the message" (Renn, 1991, p. 301), rather than an integral dimension that requires sustained analysis in its own right.

The radical tradition of research based on Jürgen Habermas' account of the mediated public sphere displays the same blind spot. Habermas is well aware that modern public culture has always been a complex mixture of information, argument, narrative, and imagery. Indeed in his seminal work, *The Structural Transformation of the Public Sphere* (1991), he points out that the political public sphere centered around the press and public debate grew out of a wider cultural sphere grounded in popular literature, theater, and display. Yet because he saw the organization of argument as the cornerstone of democratic politics, he had little interest in the communicative or rhetorical role of imagery. This is a major limitation because it is largely within the cultural public sphere that people find the symbolic resources to deal with "the chronic and persistent problems of life" (McGuigan, 2000, p. 5)—trust, betrayal, responsibility, and death. Public communication clearly involves both information transfer and argument. However, since the beginnings of the modern popular press, it has also entailed telling stories and mobilizing images, initially through cartoons and engravings and later through photojournalism and photomontage.

A recent survey of news and feature pages in a representative cross-section of British national newspapers found that over half (52%) carried at least one photograph (Huxford, 2001). Tabloid titles are even more image-intensive, with over two thirds (68%) of the total space on the front page of *The Sun*, the country's best-selling title, being taken up with photographs and headlines (Tirohl, 2000). As Roland Barthes demonstrated so forcefully in his celebrated interpretation of a *Paris Match* cover photograph showing a young black man saluting the French flag at the height of the Algerian War of Independence, images are a particularly effective way to ground interpretive frames. By giving abstract ideas (such as patriotism or imperial unity) a human face or connecting them to immediately recognizable scenes and objects, they offer safe anchorages in the windswept seas of political ideology. This search for familiar imaginative habitats is intensified when issues are surrounding by uncertainties and potential risks as they are with GM foods.

Both information technology and genetic engineering erode a sense of safety by blurring the boundaries that provide the essential grid references we use to map the world and hold it stable. Debates around artificial intelligence and visions of cyborgs that incorporate computer-controlled capacities into the human body pose fundamental challenges to our established understandings of the differences between men and machines. Similarly, producing a tomato by splicing genetic material from a fish into a familiar vegetable (as with the Flavr Savr) radically destabilizes previous certainties about the line dividing the animal and vegetable kingdoms. At the same time, the ecological models that underpin much opposition to GM foods continually point to the hidden connections between everyday actions and objects and large-scale environmental shifts. "Menace springs up behind the mundane" (Myerson, 2001, p. 69). It was precisely because visual rhetoric is able to establish such economic and vivid connections between familiar images and deep-seated concerns that it became one of the weapons of choice among critics of GM foods in 1999.

Popular visual commentary in the British press can draw on a vigorous tradition of political caricature, which runs in an unbroken line from Gilray and Richardson's savage lampoons on the corruptions of power, posted in the coffee shop windows of Regency London down to today's leading cartoonists. It also borrows extensively from the techniques of photomontage pioneered in the 1930s in the fight against Fascism.

In February 1999, the tabloid *Daily Mirror*, traditionally supportive of the Labor Party, published a front page dominated by a photomontage of Tony Blair as Frankenstein's monster. By deftly combining popular fears about the undeclared effects of "Frankenstein Foods," with suspicions that the prime minister had become a creature of the major biotech corporations, this resonant, witty image spoke powerfully for the heartland of "old" Labor against the public relations "spin" of Tony Blair's New Labor. The argument that the prime minister had become a mouthpiece for corporate interests was pursued in portraits showing his head entirely composed of vegetables—a device originally pioneered by the 16th-century Venetian painter Archimboldo. These images also played on the popular use of *vegetable* to describe someone who has lost the capacity for independent thought and speech through brain damage or other severe disability. The country's leading satirical magazine, *Private Eye*, led the way with a front cover showing a vegetable face with Blair's characteristic smile, declaring that, "There is absolutely no danger at all." This image achieved increased currency when it was reproduced to illustrate a major news feature on the GM debate published on February 21, 2001, in the major weekly broadsheet, *The Sunday Times*.

This basic idea had already been taken a stage further 4 days earlier by the cartoonist Martin Rowson of *The Guardian* with a drawing showing Lord Sainsbury, the Minister of Science, standing in front of a heap of giant vegetables saying, "Now that's what I call freedom from choice." By the time this cartoon appeared, Lord Sainsbury had already become the target of a sustained press campaign arguing that his position as chairman of the supermarket group J Sainsbury plc and as a former major investor in genetic engineering made it impossible for him to be impartial in debates on the issue.

A tomato in the front row has Blair's face and is saying, mimicking his characteristic hesitations, "It's . . . er, y'know . . . PERFECTLY SAFE! Hi!"

This image presents Tony Blair as having been modified to ensure his support. In other representations, this same metaphor is used to critique the prime minister's own efforts to suppress dissent within his own party. The Rowson Cartoon of February 16, 2001, for example, shows him standing in front of serried ranks of smiling carrots all wearing rosettes bearing the slogan "Modified Labor." The single exception is the solitary figure of Ken Livingstone, who at the time was locked in a bitter battle over nominations for the mayoral election in London. He is holding up a placard announcing "Me for Mayor." In the foreground, Blair's press secretary is whispering to him, "Don't look now boss, but we seem to have neglected to genetically modify one of the vegetables." This metaphor of *modification* was taken up by other cartoonists, including Tim in *The Independent* of May 18, 2001, who showed a smiling Blair declaring, "We think GM vegetables are so safe we've given some of them cabinet jobs!"

These images of politicians reduced to a vegetative state were not simply a witty comment on attempts to direct public debate. They also touched on deeper fears of mutation rooted in the cumulative history of nuclear power. From the outset, the promise of cheap, clean domestic energy was overshadowed by the mushroom cloud over Hiroshima, the fear of a nuclear strike from the communist East, and the proliferating images of mutation promoted by science fiction films. In the early 1980s, the U.S. Food and Drug Administration had begun licensing the use of radiation to kill bacteria in foodstuff and prolong their shelf life. The ensuing struggle over the possible risks of this process established a direct link between the known impact of exposure to radiation as a result of nuclear blasts or accidents and the possible effects of eating irradiated foods. Yet the connections between nuclear technologies and biotechnologies also operate at deeper levels, as Monsanto's CEO, Bob Shapiro, recognized. Reflecting on the popular opposition to the company's push to promote GM technologies, he told a journalist,

> When people hear about biotech, about how it's tinkering with the very essence of life the immediate association is to nuclear science. It's dawned on them that we have probed the mysteries of the universe down to the atomic level, and look what happens: Boom! You kill millions of people, you poison the air. (Herrera, 2000, pp. 162–164)

Tony Blair inadvertently suggested this link in a TV interview in June 1999. Describing his own initial reaction to GM foods, he remembered thinking of Dr. Strangelove, the demented pronuclear scientist in Stanley Kubrick's film:

> The first time I heard about genetic modification . . . You think, my goodness, what on earth is going on here? You think of Dr. Strangelove. (quoted in Watt, 1999, p. 1)

These associations between nuclear dangers and GM risks were frequently reinforced in popular press iconography. The *Independent on Sunday*, which was consistently critical of the government's positive stance on GM foods, chose the familiar hazard warning sign at nuclear plants as the logo for its "Stop GM foods" campaign and continually returned to the association. On May 2, 1999, it carried a story headed: "Ministers told to study GM food cancer risks." The text made no mention of nuclear risks and framed the announcement in relation to a similar research initiative on the possible impact of BSE on humans. The cartoon that accompanied the item, however, showed a man standing in the hallway of his house dressed from head to foot in the kind of protective suit worn in nuclear plants and familiar from numerous thrillers and science fiction films. He is saying to his wife, "I'm just popping down to the greengrocers."

This image of protective clothing gained extensive coverage 2 months later when, as mentioned earlier, Greenpeace released photographs and a videotape of its members uprooting a trail planting of GM crops wearing what they described as decontamination suits. These costumes were similar to those worn by the safety teams who had reentered the nuclear reactor at Chernobyl after it had discharged contamination across a huge area of Northwestern Europe. Following the accident images of clean-up teams, wearing these suits became a stock element in popular iconography, easily reactivated. In April 1996, for example, a decade after the disaster, *The Guardian* carried a story headed, "Deadly Shadow Hangs Over Europe." The accompanying photograph, spread over seven columns, showed a long line of volunteers in protective suits waiting to enter the devastated plant (see Boholm, 1998, p. 141). The suits were also reminiscent of the clothing worn by forensic teams working at the scenes of disasters or violent crimes frequently featured on news bulletins. It was a visual propaganda coup, a perfectly managed image event.

Beneath these immediate connotations, however, these images also detonated a third level of association rooted in the idea that both nuclear power and genetic modification might be capable of producing the kind of desolation prophesied in the Bible. In his analysis of the press photographs printed to illustrate stories marking the 10th anniversary of Chernobyl, Boholm (1998) notes how textural accounts of the event as an "accident," accountable in terms of science, were consistently challenged by accompanying photographs, which portrayed it as "an event preordained by divine forces striking at sinful humanity" (p. 139).

The move from assumptions of fate to calculations of probability is generally seen as one of the defining breaks with religious schemas that ushered in a modern age. As commentators from Max Weber onward have noted, however, the world may have been secularized by the rise of science, but the spiritual vacuum created by the onward march of instrumental reason has left an empty symbolic space waiting to be filled with religious imagery that speaks to deep-seated fears over matters of life and death (see Murdock, 1997). Anti-GM activists understood this very well and were photographed in trial fields and on street demonstrations dressed as the Grim Reaper, the harbinger of death in traditional Christian iconography. The most concerted attempt to mobilize religious frames against GM foods, however, came from the heir to the throne, Prince Charles. As both a passionate advocate of organic farming and the future head of the state church, he was particularly well placed to knit together popular notions of national tradition and diffuse misgivings about the possible long-term impact of GM foods.

In May 2000, Prince Charles was invited to give one of the annual Reith lectures, a prestigious series of radio talks honoring the BBC's first director general, John Reith. His argument, which had been posted some time before on his personal website, deftly combined a critique of scientific overreaching with a defense of traditional farming practices:

> Mixing genetic material from species that cannot breed naturally takes us into areas that should be left to God. We should not be meddling with the building blocks of life in this way.... I am not convinced that we know enough about the long-term consequences for human health and the environment of releasing plants (or, heaven forbid, animals) bred in this way.... Wouldn't it be better to concentrate instead on the sustainable techniques which can double or treble the yields from traditional farming methods. (Prince of Wales, 1999)

At the top of the web page on which this text appeared was a photograph. It showed the Prince standing in a sunlit field, with woodland behind, leaning on a traditionally carved shepherd's crook looking resolutely but benignly into the camera. It cleverly combined connotations of national heritage and continuity with echoes of countless images of Jesus as the good shepherd guiding his flock onto the path of righteousness. This image was reprinted frequently in the daily press as the argument over Charles' remarks gathered momentum, often generating tensions between the argument in the text and the connotations carried by the image. At the time of his Reith lecture, for example, *The Guardian* of May 20th carried a full-page spread that was generally skeptical of both the Prince's arguments and his claims to speak for public opinion. The main story, which was headed, "Prince's luxury lifestyle conflicts with image of ascetic and reluctant royal," pointed out that although he had composed his lecture on retreat in a Greek monastery, the prince had arrived there "on board the third biggest luxury yacht in the world . . . plaything of his friend, the elderly Greek shipping tycoon, John Latsis. . . . For a man concerned with sustainable development this emblem of conspicuous consumption seemed to jar" (Meek, 2000, p. 3). The story occupied three columns. The photograph, reproduced in full color, was spread across six columns and dominated the page layout.

THE POLITICS OF REPRESENTATION REVISITED

In English, *representation* refers to systems of social delegation through which individuals and groups claim to speak on behalf of particular constituencies and to the array of cultural forms through which key ideas and values are expressed. The struggle against GM foods outlined here provides an instructive case study in the politics of representation in both these senses. It shows a range of nongovernmental organizations seizing the political initiative and deploying resonant linguistic and visual signs, from "Frankenfoods" to "decontamination" suits, to crystallize public anxieties and challenge the concerted effort made by the biotechnology companies and the government to extend the commercial planting and sale of GM foods at the earliest possible opportunity.

The conservative French commentator, Gustav Le Bon, was convinced that he could already see this new politics of representation emerging in 1895. He was writing against the background of the Paris Commune, when the people of Paris had set up their own forms of government and waged a fierce "war of images" against their detractors, mobilizing every possible popular visual form from cartoons and caricatures to vivid posters using the new lithographic techniques (Leith, 1978). The year after his book came out, the Lumiere brothers mounted the world's first commercial cinema performance in Paris, making images even more central to popular experience. Le Bon saw these developments and concluded that the popular politics of the future would be dominated by images. "A crowd," he complained, "thinks in images, and the image itself immediately calls up a series of other images, having no logical connection with the first" (Le Bon, 1895/1969, p. 36). In common with many other writers, he believed that by abandoning the disciplines of logical sequence entailed in linguistic expression, these chains of visual association undermined rational debate and encouraged emotive responses rather than careful evaluations of relevant evidence. Senior British government figures made the same case in 1999, repeatedly criticizing the press for sensationalizing the issues and creating unnecessary public fear and concern. The analysis presented here, however, suggests another explanation for the central role of images.

In conditions of postnormal science, when doubts are continually being expressed about how far to trust scientific authority and about the human and environmental consequences of new technologies, issues of social interests and moral values become ever more salient. The relevant questions are not simply, "What do we know about the possible consequences of a particular intervention?", but, "Who will benefit from it and is it morally justifiable?" Empirical evidence cannot answer these questions. They take us beyond calculations of risk and confront us with fundamental political and moral decisions. In working our way through these choices, images and stories provide essential resources. They dramatize our dilemmas and allow us to take imaginative walks to a variety of possible destinations. They are not irrational. Rather, they are against the uncritical application of instrumental rationality, which reduces science to a tool kit for reconstructing the world.

Grounds for Action

At first sight, the headlong pursuit of instrumental rationality seems to have been defeated in the struggle over GM foods in Britain in 1999. In February 2000, almost 1 year after the row first broke, Tony Blair contributed a conciliatory article to the *Independent on Sunday*—a paper consistently critical of his position. He conceded that, "there is cause for legitimate public concern" about the "potential for harm, both in terms of human safety and in the diversity of our environment, from GM foods and crops" and sought to reassure readers that "the protection of the public and the environment is, and will remain, the Government's over-riding priority" (Blair, 2000, p. 28). Speaking to the business community, however, he employed a rather different rhetoric. Nine months later, in a speech to the European Bioscience Conference in London, he roundly condemned the sympathetic publicity given GM protesters, arguing that, "To make heroes of people who are preventing basic science taking place is wrong [and] is to substitute aggression for argument," and reaffirmed his full support for "The giants of British biotechnology," adding that "we don't intend to let our leadership fall behind" (quoted in Clark, 2000, p. 27). The government demonstrated its determination to curb the aggression of the crop protesters in practical terms with a new Terrorism Act passed in 2000. This introduced a far more capacious definition of terrorism than previously, extending it to include "the use or threat of action ... for the purpose of advancing a political, religious or ideological cause; action which involves serious violence against a person or serious damage to property." As one critic noted, "It is hard to imagine that the state could go further" (Cohen, 2001, p. 26).

These new provisions have not deterred GM activists, however, nor have the increased security measures at test sites. Over a 12-night period in August 2001, for example, protesters in Essex managed to destroy three quarters of a trial GM planting despite the installation of infrared sensors, security cameras, dog patrols, and helicopter sweeps. They evaded cameras after children found a lens sticking out of a bird box and avoided helicopter heat-seeking devices by hiding under a tractor. Three were caught and charged with damages relating to criminal trespass, but incurred fines of only 20 pence (Vidal, 2001a).

The activists' case for abandoning all GM plantings appears increasingly plausible in the light of recent scientific research. In February 2002, English Nature, the national agency responsible for environmental protection, published a report on recent Canadian experience of "gene stacking," whereby cross-pollination results in genes from separate GM varieties accumulating in volunteer plants that grow from seed spilled at harvest, making them resistant to a range of widely used herbicides. To control them, farmers have to resort to substantial applications of old herbicides with all their well-established negative effects. This finding follows hard on the heels of a report from a group of researchers at the University of California–Berkeley, announcing that they had discovered elements from genetically modified varieties in native cirollo corn crops in the Sierra Norte de Oaxaca region of Mexico, although the Mexican government banned the growing of GM maize in 1998. They speculated that the elements may have originated in GM maize imported from the United States as food aid and progressed over time through multiple pollinations. This "escape," in a region particularly known for the variety of its native corn species, has major implications for the future genetic diversity of one of the world's staple foodstuffs (Vidal, 2001b).

Both these studies underline once again the two defining features of scientific inquiry in postnormal times: the inability to fully predict and control complex chains of possible impact, and the recognition that technical innovation is inextricably bound up with questions of power and ethics. As a recent Royal Society (2002) report on GM plants noted, "the public debate about GM food must take account of wider issues that the science alone" (p. 3). The fact that a body that had previously been an enthusiastic advocate of the potential benefits of GM plants now recognizes this and has invited activist groups such as Greenpeace and Friends of the Earth to take part in its deliberations is a signal of how far the debate has moved.

At the same time, it is important to recognize that the destruction of the twin towers of the World Trade Center in New York on September 11, 2001, has altered the terms of debate in important ways. The bombing and its aftermath have dramatized the transition from industrial to information capitalism in the starkest possible terms. In the attack, three of the key symbols of the machine age—the skyscraper, the jet, and oil-fueled power—collided to terrible effect as triumphs of civil engineering were converted into sites of mass destruction. In the weeks that followed, however, attention gradually shifted from the twisted metal at Ground Zero to the risks embedded in the technologies of information and bioscience at the heart of the emerging economic system. It became all too evident that the key resources for action have become uniquely mobile. The computer networks that had carried the last desperate e-mails sent by those trapped on the upper floors of the World Trade Center had earlier carried the flows of money and information that made the attack possible. The biosciences that had helped develop the drugs stored in the hospitals that treated the survivors had also ushered in the age of bioterrorism. The postal system that delivered letters of condolence to the bereaved also carried envelopes containing anthrax.

Governments on both sides of the Atlantic Ocean have responded to the realization that biological knowledge can no longer be securely contained and the new networks of information capitalism are chronically "leaky," by launching a "global war against terrorism." Like the Cold War against the Soviet Union, this involves a protracted struggle for political and moral ascendancy fought across multiple theaters of conflict. The new rhetoric of an "axis of evil" draws on the powerful connotations established by earlier characterizations of the "Evil Empire" of Soviet domination and the "Axis Powers" of World War II. In a war, only one question is seen to matter: "Are you with us or against us?" Dissenting voices are easily dismissed as the "enemy within." In this situation, it is more important than ever that scholars continue to investigate how biotechnologies and information systems are being deployed in the emerging world order and persist in asking these awkward questions: Where are we going? Who gains and who loses, by which mechanisms of power? Is it desirable? What can be done?

References

A thousand billion suns. (1995, May 27). *The Economist, 335*(7916), 73–75.

American Association of University Professors. (2001, May–June). Statement on corporate funding of academic research. *Academe, 87*(3), 68–70.

Anderson, B. (1983). *Imagined communities: Reflections on the origin and spread of nationalism.* London: Verso.

Anderson, L. (1999). *Genetic engineering, food, and our environment: A brief guide.* Dartington: Green Books.

Andersson, S. (1991). Bowers on the savana: Display courts and mate choice in a lekking widowbird. *Behavioral Ecology, 2*(3), 210–218.

Antonelli, C. (1992). *The economics of information networks.* Amsterdam: North Holland.

Aoki, K. (1996). (Intellectual) property and sovereignty: Notes toward a cultural geography of authorship. *Stanford Law Review, 48*(5), 1293–1355.

Aoki, K. (1999). Neo-colonialism, anticommons property and biopiracy in the (not-so-brave) new world order of international intellectual property protection. *Global Legal Studies Journal, 6*(1) (http://www.law.indiana.edu/glsj/vol6/no1/aoki.html, consulted February 18, 2000).

Appadurai, A. (1993). Consumption, duration, and history. In D. Palumbo-Liu & H. Ulrich Gumbrecht (Eds.), *Streams of cultural capital: Transnational cultural studies* (pp. 23–46). Stanford: Stanford University Press.

Arrow, K. (1996). The economics of information: An exposition. *Empirica, 23*(2), 119–128.

Arthur, C. (1999, June 16). We don't need GM foods, Meacher tells Commons. *The Independent,* p. 8.

Bagwell, K., & Ramey, G. (1993, Summer). Advertising as information: Matching products to buyers. *Journal of Economics and Management Strategy, 2*(2), 199–243.

Bailey, R. (1988, June 27). Ministry of fear: Jeremy Rifkins' war against biotechnology. *Fortune, 141*(14), 138–140.

Baker, J. P. (2000). Immunization and the American way: 4 childhood vaccines. *American Journal of Public Health, 90,* 199–207.

Barnes, D. E., & Bero, L. A. (1998). Why review articles on the health effects of passive smoking reach different conclusions. *Journal of the American Medical Association, 279*(19), 1566–1570.

Baudrillard, J. (1983). *Simulations.* New York: Semiotexts.

Baudrillard, J. (2000). *The vital illusion* (J. Witwer, Trans.). New York: Columbia University Press.

Beachy, R. N. (1988). Reflections of an industry-supported university scientist. In P. DeForest, M. S. Frankel, J. S. Poindexter, & V. Weil (Eds.), *Biotechnology: Professional issues and social concerns* (pp. 28–33). Washington, DC: American Association for the Advancement of Science.

Beaumont, R. (1984). *War, chaos, and history.* New York: Praeger.

Beck, U. (1992). *Risk society: Towards a new modernity.* London: Sage.

Beck, U. (1999). *World risk society.* Malden, MA: Polity.

Bell, D. (1973). *The coming of post-industrial society: A venture in social forecasting.* New York: Basic Books.

Beller, J. L. (1996). Desiring the involuntary: Machinic assemblage and transnationalism in Deleuze and *Robocop 2.* In R. Wilson & W. Dissanayake (Eds.), *Global/local: Cultural production and the transnational imaginary* (pp. 193–218). Durham, NC: Duke University Press.

Benowitz, S. (1996, April 1). Progress impeded? *The Scientist, 10*(7), 1 (http://www.the-scientist.com/yr1996/apr/mum_960401.html, consulted July 11, 2001).

Benson, J., & Shaw, G. (Eds.). (1992). *The evolution of retail systems, 1800–1914.* Leicester: Leicester University Press.

Bent, S. A., Schwaab, R. L., Conlin, D. G., & Jeffery, D. D. (1987). *Intellectual property rights in biotechnology worldwide.* New York: Stockton.

Bergquist, L. (1996, August 22). BGH still hasn't gained favor among state's dairy farmers. *Milwaukee Journal Sentinel,* p. 1.

Berle, P. A. (1988, January). Do it right this time. *Audubon, 90,* 13.

Best, S., & Kellner, D. (1997). *The postmodern turn.* New York: Guilford.

Best, S., & Kellner, D. (2001). *The postmodern adventure: Science, technology, and cultural studies at the Third Millennium.* New York: Guilford.

Bettelheim, A. (2001, May 5). Senate panel considers a bill to prevent human cloning by banning a lab technique. *CQ Weekly, 59,* 1019.

Biorhythms. (1993, July 3). *The Economist, 328*(7818), 74.

Birke, L. I. A., & Hubbard, R. (1995). *Reinventing biology: Respect for life and the creation of knowledge.* Bloomington: Indiana University Press.

Black, R. (2003, February 16). *Bioterror fears muzzle open science.* BBC News, online.

Blair, T. (2000, February 27). The key to GM is it potential both for harm and good. *The Independent on Sunday,* p. 28.

Blumenstyk, G. (1998, December 4). Berkeley joins Swiss company in controversial technology-transfer pact. *The Chronicle of Higher Education,* p. A38.

Blumenthal, D., Campbell, E. G., Anderson, M. S., Causino, N., & Louis, K. S. (1997, April 16). Withholding research results in academic life science: Evidence from a national survey of faculty. *Journal of the American Medical Association, 277,* 1224–1228 (http://www.ama-assn.org/sci-pubs/journals/archive/jama/feature06.html, consulted July 11, 2001).

Blythman, J. (1999, February 7). Why we must end this now. *The Independent on Sunday,* p. 8.

Boholm, A. (1998). Visual images and risk messages: Commemorating Chernobyl. *Risk Decision and Policy, 3*(2), 125–143.

Borges, J. L. (1962). The library of Babel. In *Labyrinths: Selected stories and other writings* (pp. 51–58). New York: New Directions.

Boseley, S. (2002, February 7). Scandal of scientists who take drug money for papers ghostwritten by drug companies. *The Guardian,* p. 4.

Bowler, I. R. (Ed.). (1992). *The geography of agriculture in developed market economies.* New York: Longman Scientific and Technical.

Boyle, J. (1996). *Shamans, software and spleens: Law and the construction of the information society.* Cambridge, MA: Harvard University Press.

Boyle, J. (1997). A politics of intellectual property: Environmentalism for the Net? *Duke Law Journal, 47*(1), 87–116.

Brainerd, J. (1998, September 26). Chaucer's descendants: Evolutionary biologists help trace the ancestry of a classic. *Science News, 154*(13), 200.

Braman, S. (1993). Harmonization of systems: The third stage of the information society. *Journal of Communication, 43*(3), 133–140.

Braman, S. (1995). Horizons of the state: Information policy and power. *Journal of Communication, 45*(4), 4–24.

Braman, S. (2002a). Informational meta-technologies, international relations and genetic power: The case of biotechnologies. In J. P. Singh & J. Rosenau (Eds.), *Information technologies and global politics: The changing scope of power and governance* (pp. 91–112). Albany, NY: State University of New York Press.

Braman, S. (2002b). Posthuman law: Information policy and the machinic world. *First Monday* (http://www.firstmonday.org/issues/issue7_12/braman/index.html).

Braman, S., & Lynch, S. (2003). Advantage ISP: Terms of service as media law. In L. Cranor & S. Wildman (Eds.), *Rethinking rights and regulations: Institutional responses to new communication technologies* (pp. 249–278). Cambridge, MA: MIT Press.

Brandenberger, A. M., & Nalebuff, B. J. (1996). *Co-opetition.* New York: Doubleday.

Branscomb, L. (Ed.). (1993). *Empowering technology: Implementing a US policy.* Cambridge: MIT Press.

Braudel, F. (1977). *Afterthoughts on material civilization and capitalism* (P. M. Ranum, Trans.). Baltimore: The Johns Hopkins University Press.

Brown, P. (1999, February 19). Meacher puts GM crops on hold. *The Guardian,* p. 1.

Bud, R. (1993). *The uses of life: A history of biotechnology.* Cambridge: Cambridge University Press.

Bugged. (1994, December 3). *The Economist, 333*(7892), 58.

Bull, A. T., Holt, G., & Lilly, M. D. (1982). *Biotechnology: International trends and perspectives.* Paris: Organization for Economic Cooperation and Development.

Bullard, C. W. (1988). Management and control of modern technologies. *Technology in Society, 10,* 205–232.

Burch, K. (1995). Intellectual property rights and the culture of global liberalism. *Science Communication, 17*(2), 214–232.

Burress, C. (1998, November 24). UC finalizes pioneering research deal with biotech firm. *The San Francisco Chronicle,* p. A17.

Busch, L., Lacy, W. B., Burkhardt, J., & Lacy, L. R. (1991). *Plants, power and profit: Social, economic, and ethical consequences of the new biotechnologies.* Oxford: Basic Blackwell.

Cambrosio, A., Keating, P., & Mackenzie, M. (1990). Scientific practice in the courtroom: The construction of sociotechnical identities in a biotechnology patent dispute. *Social Problems, 37*(3), 275–293.

Camerer, C. (1988). Gifts as economic signals and social symbols. *American Journal of Sociology, 94*(Suppl.), S180–S214.

Caron, J. A. (1988). "Biology" in the life sciences: A historiographical contribution. *History of Science, 26,* 223–268.

Carr, G. (2000, July 1). A survey of the human genome. *The Economist* (Suppl.).

Carrol, R. (1999, February 19). Gene crops could spell extinction for birds. *The Guardian,* p. 9.

Carving up Europe's drugs industry: Pharmaceuticals. (1995, August 26). *The Economist, 336*(7929), 57.

Castells, M. (2001). *The internet galaxy: Reflections on the internet, business, and society.* Oxford: Oxford University Press.

Chandler, A. D., Jr. (1977). *The visible hand: The managerial revolution in American business.* Cambridge: Belknap Press.

Chandler, A. D., Jr., & Cortada, J. W. (Eds.). (2000). *A nation transformed by information: How information has shaped the United States from colonial times to the present.* Cambridge: Harvard University Press.

Chanley, V. A., Rudolph, T. J., & Rahn, W. M. (2000). The origins and consequences of public trust in government: A time series analysis. *The Public Opinion Quarterly, 64*(3), 239–256.

Clark, A. (2000, November 18). Blair derides "anti-science." *The Guardian,* p. 27.

Clark, N., & Juma, C. (1991). *Biotechnology for sustainable development: Policy options for developing countries.* Nairobi: African Center for Technology Studies.

Clarke, A., & Montini, T. (1993). The many faces of RU486: Tales of situated knowledges and technological contestations. *Science, Technology, and Human Values, 18*(1), 42–78.

Cleveland, D. A., Soleri, D., & Smith, S. E. (1994). Do folk crop varieties have a role in sustainable agriculture? *Bioscience, 44*(11), 740–751.

Cohen, N. (2001, October 1). Make the computers work first! *New Statesman,* pp. 24–26.

Cohn, V. (1977, July 18). Scientists now downplay risks of genetic research; Gene research risks now downplayed. *Washington Post,* p. A1.

Commons (1924/1959). *Legal foundations of capitalism.* New York: Macmillan.

Connor, S. (1999a, May 19). Findings of GM foods scientist defective, says Royal Society. *The Independent,* p. 10.

Connor, S. (1999b, May 20). Modified pollen kills threatened butterfly. *The Independent,* p. 1.

Cotts, C. (2000, December 5). Does Fox slant the news? *The Village Voice,* p. 39 (issues/issue6_1/kahin, January 10, 2001).

Crespi, R. S. (2000). An analysis of moral issues affecting patenting inventions in the life sciences: A European perspective. *Science & Engineering Ethics, 6*(2), 157–180.

Cribbet, J. E. (1986). Concepts in transition: The search for a new definition of property. *University of Illinois Law Review, 1,* 1–42.

Crichton, M. (1969). *The Andromeda strain.* New York: Random House.

Crosby, A. W., Jr. (1972). *The Columbian exchange: Biological and cultural consequences of 1492.* Westport, CT: Greenwood.

Crosby, A. W., Jr. (1994). *Germs, seeds & animals: Studies in ecological history.* Armonk, NY: M. E. Sharpe.

Crucible Group, The. (1994). *People, pants, and patents: The impact of intellectual property on trade, plant diversity, and rural society.* Ottawa: IDRC.

Davis, L. J. (1983). *Factual fictions: The origins of the English novel.* New York: Columbia University Press.

Dawkins, R. (1976). *The selfish gene.* Oxford: Oxford University Press.

Dawkins, R. (1989). *The selfish gene* (2nd ed.). Oxford: Oxford University Press.

Dawkins, R. (1993). Viruses of the mind. In B. Dahlbom (Ed.), *Dennett and his critics* (pp. 13–27). Oxford: Blackwell.

Dawkins, R. (1998). *Unweaving the rainbow: Science, delusion and the appetite for wonder.* Boston: Houghton-Mifflin.

Dawson, A. C. (1998). The intellectual commons: A rationale for regulation. *Prometheus, 16*(3), 275–289.

Dechema. (1982). *Biotechnology in Europe* (FAST Occasional Paper #59, XII-701-82). Brussels: Organization for Economic Cooperation and Development.

DeForest, P. (1988). Industry support of university biotechnology research: An introduction to the issues and perspectives. In P. DeForest, M. S. Frankel, J. S. Poindexter, & V. Weil (Eds.), *Biotechnology: Professional issues and social concerns* (pp. 5–11). Washington, DC: American Association for the Advancement of Science.

DeForest, P., Frankel, M. S., Poindexter, J. S., & Weil, V. (Eds.). (1988). *Biotechnology: Professional issues and social concerns.* Washington, DC: American Association for the Advancement of Science.

de Freitas Filho, A., Paez, M. L. D., & Goedert, W. J. (2002). Strategic planning in public R&D organizations for agribusiness. *Technological Forecasting and Social Change, 69*(8), 833–847.

DeLanda, M. (1991). *War in the age of intelligent machines.* New York: Zone.

Delaney, E. J. (1993). Technology search and firm bounds in biotechnology: New firms as agents of change. *Growth and Change, 24*(2), 206–228.

Delucca, M. (1999). *Image politics: The new rhetoric of environmental activism.* London: Routledge.

Dennett, D. C. (1995). *Darwin's dangerous idea: Evolution and the meanings of life.* New York: Simon & Schuster.

Der Derian, J. (1990). The (s)pace of international relations: Simulation, surveillance, and speed. *International Studies Quarterly, 34,* 295–310.

Desrosiéres, A. (1998). *The politics of large numbers: A history of statistical reasoning.* Cambridge: Harvard University Press.

Dewdney, C. (1998). *Last flesh: Life in the transhuman era.* San Francisco: Harper.

Dezalay, Y., & Garth, B. G. (1996). *Dealing in virtue: International commercial arbitration and the construction of a transnational legal order.* Chicago: University of Chicago Press.

Diamond v Chakrabarty, 447 US 303 (1980).

Dickey, C. (2001, April). I love my glow bunny. *Wired,* pp. 88–100.

Douglas, M. (1992). Risk and justice. In *Risk and blame: Essays in cultural theory* (pp. 22–37). New York: Routledge.

Douglas, M., & Wildavsky, A. (1982). *Risk and culture: An essay on the selection of technical and environmental dangers.* Berkeley: University of California Press.

Douthitt, R. A. (1995). Consumer risk perception and recombinant bovine growth hormone: The case for labeling dairy products made from untreated herd milk. *Journal of Public Policy & Marketing, 4*(2), 328–330.

Ducor, P. G. (1998). *Patenting the recombinant products of biotechnology and other molecules.* London: Kluwer Law International.

Duncan, M. R. (1989). US agriculture: Hard realities and new opportunities. In M. Drabenstott (Ed.), *Positioning agriculture for the 1990s: A new decade of change* (pp. 3–28). Washington, DC: National Planning Association.

Durant, J., & Lindsey, N. (2000). *The "Great GM Food Debate": A survey of media coverage in the first half of 1999.* House of Commons, Parliamentary Office of Science and Technology (education_integrity.htm).

Dyson, G. (1997). *Darwin among the machines: The evolution of global intelligence.* Reading, MA: Perseus.

Eccles, L. (2001, April 2). ISP plans to turn subscriber base into supercomputer network. *Electronic Design, 49*(7), 34.

The Economist. (2000, July 1). Survey: The human genome, Special section, pp. 1–16.

The Economist. (2001a, May 19). Outrageous fortune, pp. 77–78.

The Economist. (2001b, September 22). Drugs ex machina, Technology Quarterly section, pp. 30–32.

Ehrlich, P., Ehrlich, A., & Daily, G. C. (1993). Food security, population and environment. *Population and Development Review, 19*(1), 1–32.

Eisenberg, R. S. (1996). Intellectual property at the public-private divide: The case of large-scale cDNA Sequencing. *University of Chicago Law School Roundtable, 3,* 557–573.

Eisenstein, E. L. (1979). *The printing press as an agent of change: Communications and cultural transformations in early-modern Europe.* Cambridge: Cambridge University Press.

Elias, T. D. (1998, December 4). Pies fly over university's deal on genetically altered foods. *The Washington Times,* p. A4.

Elichirigoity, F. (1999). *Planet management: Limits to growth, computer simulation, and the emergence of global spaces.* Evanston, IL: Northwestern University Press.

Eliot, T. S. (1934). The rock. In *The waste land and other poems* (pp. 80–88). New York: Harcourt, Brace.

Ellul, J. (1964). *The technological society* (J. Wilkinson, Trans.). New York: Vintage Books.

English Nature. (2002). *Gene-stacking in herbicide-tolerant oilseed rape: Lessons from the North American experience* (Research Report Number 443). London: Author.

Enzensberger, H. M. (1974). *The consciousness industry: On literature, politics and the media.* New York: Seabury Press (eon.law.harvard.edu/property/history.html, accessed January 26, 2001).

Erecky, K. (1919). *Biotechnologie der Fleisch-, Fett- und Milcherzeugung im landwirtschaftlichten Großbetriebe.* Berlin: Paul Parey.

ESRC Global Environmental Change Programme. (1999). *The politics of GM food: Risk, science and public trust* (Special Briefing Paper No. 5). Sussex: University of Sussex.

Evans, W., & Priest, S. (1995). Science content and social contest. *Public Understanding of Science, 4,* 327–340.

Evolution in a test tube. (1993, August 7). *Science News, 144*(6), 90–92.

Exterminate, exterminate. (2003, March 22). *The Economist,* pp. 73–74.

Falaschi, A., & Tzotzos, G. T. (1993). Preface. In G. T. Tzotzos (Ed.), *Biotechnology R&D trends: Science policy for development, Annals of the New York Academy of Sciences, 700* (pp. ix–x). New York: New York Academy of Sciences.

Falk, R. (2000). Humane governance for the world: Reviving the quest. *Review of International Political Economy, 7*(2), 317–334.

Farrel, J. (1987). Cheap talk, coordination, and entry. *Rand Journal of Economics, 18*(1), 34–39.

Farrel, J., & Rabin, M. (1993). Meaning and credibility in cheap-talk games. *Games and Economic Behavior, 5,* 514–531.

Farrel, J., & Rabin, M. (1996, Summer). Cheap talk. *Journal of Economic Perspectives, 10*(3), 103–118.

Finn, S., & Roberts, D. F. (1984). Source, destination, and entropy: Reassessing the role of information theory in communication research. *Communication Research, 11*(4), 453–476.

Fisher, W. W. III. (1999). The growth of intellectual property: A history of the ownership of ideas in the United States (http://eon.law.harvard.edu/property/history.html, January 26, 2001).

Fleming, D. (1968). Emigre physicists and the biological revolution. *Perspectives in American History, 2,* 152–189.

Flor, A. G. (1993). The informatization of agriculture. *Asian Journal of Communication, 3*(2), 94–103.

Foucault, M. (1972). *The archaeology of knowledge & the discourse on language* (A. M. Sheridan Smith, Trans.). New York: Pantheon.

Foucault, M. (1978). *The history of sexuality.* New York: Pantheon.

Fox, M. W. (1999). *Beyond evolution: The genetically altered future of plants, animals, the earth—and humans.* New York: Lyons.

Frankel, O. H. (1988). Genetic resources: Evolutionary and social responsibilities. In J. Kloppenburg, Jr., & D. L. Kleinman (Eds.), *Seeds and sovereignty: The use and control of plant genetic resources* (pp. 19–48). Durham, NC: Duke University Press.

Friedman, S. M., Dunwoody, S., & Rogers, C. L. (1999). *Communicating uncertainty: Media coverage of new and controversial science.* Mahwah, NJ: Lawrence Erlbaum Associates.

Fukuyama, F. (2002). *Our posthuman future: Consequences of the biotechnology revolution.* New York: Farrar, Straus, Giroux.

Fuller, R. B. (1975). *Synergetics: Explorations in the geometry of thinking.* New York: Macmillan.

Funtowicz, S. O., & Ravetz, J. R. (1992). The good, the true and the post-modern. *Futures, 24,* 739–752.

Gandy, O. (1982). *Beyond agenda setting: Information subsidies and public policy.* Norwood, NJ: Ablex.

Garber, K. (2000, March 3). Homestead 2000: The Genome. *Signals* (http://www.signalsmag.com, March 13, 2001).

Garland, M. J. (1999). Experts and the public: A needed partnership for genetic policy. *Public Understanding of Science, 8,* 241–254.

Gaskell, G., Ten Eyck, T., Einsiedel, E., Priest, S., Allum, N., & Torgersen, H. (2001). Troubled waters: The Atlantic divide on biotechnology policy. In G. Gaskel & M. W. Bauer (Eds.), *Biotechnology, 1996–2000: The years of controversy* (pp. 96–105). London: Science Museum.

Geller, P. E. (1994). Must copyright be forever caught between marketplace and authorship norms? In B. Sherman & A. Strowel (Eds.), *Of authors and origins* (pp. 159–201). Oxford: Clarendon.

Gibas, C., & Jambeck, P. (2001). *Developing bioinformatics computer skills.* Sebastopol, CA: O'Reilly and Associates.

Gibbs, W. W. (1996, November). The price of silence: Does profit-minded secrecy retard scientific progress? *Scientific American, 275*(5), 15–16.

Giddens, A. (1991). *Modernity and self-identity.* Stanford, CA: Stanford University Press.

Glanz, J. (2001, June 19). The Web as dictator of scientific fashion. *New York Times,* pp. D1–D2.

Golden, J. M. (2001). Biotechnology, technology policy and patentability: Natural products and invention in the American system. *Emory Law Journal, 50*(1), 101–191.

Goldstein, J. (1988). Ideas, institutions, and American trade policy. *International Organization, 42*(1), 179–217.

Gollier, C. (2001, October). Should we beware of the precautionary principle? *Economic Policy: A European Forum, 33,* 303–327.

Goodman, D., Sorj, B., & Wilkinson, J. (1987). *From farming to biotechnology: A theory of agroindustrial development.* Oxford: Basil Blackwell.

Goodwin, B. C. (2001). *How the leopard changed its spots: The evolution of complexity.* Princeton: Princeton University Press.

Goonatilake, S. (1991). *The evolution of information: Lineages in gene, culture, and artefact.* New York: Pinter.

Gottweis, H. (1995). German politics of genetic engineering and its deconstruction. *Social Studies of Science, 25,* 195–235.

Gould, S. J. (1997). *Dinosaur in a haystack: Reflections in natural history.* New York: Crown.

Government Millennium Centre. (1999). Home Page (www.millennium-centre.gov.uk).

Government Millennium Centre. (2000). Bulletins issued at 06.600 GMT on 1 January 2000 (www.millenium-centre.gov.uk/bulletins/000101_0600.htm).

Grabher, G. (1993). *The embedded firm: On the socioeconomics of industrial networks.* New York: Routledge.

Grafen, A. (1990a). Biological signals as handicaps. *Journal of Theoretical Biology, 144,* 517–546.

Grafen, A. (1990b). Sexual selection unhandicapped by the fisher process. *Journal of Theoretical Biology, 144,* 475–516.

Greaves, T. (Ed.). (1994). *Intellectual property rights for indigenous peoples: A sourcebook.* Oklahoma City, OK: Society for Applied Anthropology.

Greek, R., & Greek, J. S. (2000). *Sacred cows and golden geese: The human cost of experiments on animals.* New York: Continuum.

Greene, R. (1993a, April 1). FDA Panel gives green light to BGH. *Capital Times,* p. B2.

Greene, R. (1993b, May 5). FDA panels to look at BHG labeling. *Capital Times,* p. B6.

The Guardian. (1999, February 12). Top researchers back suspended lab whistle blower, p. 7.

The Guardian. (2000, November 15). Patenting life: Special report (Suppl.).

Guattari, F. (1992). Regimes, pathways, subjects. In J. Crary & S. Kwinter (Eds.), *Incorporations* (pp. 16–35). New York: Zone.

Guattari, F. (1995). *Chaosmosis: An ethico-aesthetic paradigm* (J. Pefanis & P. Bains, Trans.). Bloomington, IN: Indiana University Press.

Habermas, J. (1984/1987). *The theory of communicative action.* Boston: Beacon.

Habermas, J. (1991). *The structural transformation of the public sphere: An inquiry into a category of bourgeois society.* Cambridge, MA: The MIT Press.

Hadwiger, D. F. (1982). *The politics of agricultural research*. Lincoln: University of Nebraska Press.

Haines, R. C., & Joyce, F. E. (1987). *Monitoring and management of renewable resources: The use of remote sensing—Case study for the UK and West Germany* (FAST Internal Paper 191, XII-591-87). Brussels: Organization for Economic Cooperation and Development.

Hall, S. (1980). Encoding/decoding. In S. Hall et al. (Eds.), *Culture, media, language* (pp. 128–139). London: Hutchinson.

Hall, S. (1999, March 1). Monsanto ads condemned. *The Guardian*, p. 5.

Hallin, D. (1989). *The uncensored war*. Berkeley: University of California Press.

Hamstra, I. A. (1998). *Public opinion about biotechnology*. The Hague, Netherlands: European Federation of Biotechnology.

Haraway, D. (1976). *Crystals, fabrics, and fields: Metaphors of organicism in twentieth-century developmental biology*. New Haven, CT: Yale University Press.

Haraway, D. (1997). *Modest witness@second millennium. Female meets oncomouse*. New York: Routledge.

Hardin, G. (1968/1993). The tragedy of the commons. *Science, 162*, 1243–1248. Reprinted in R. Dorfman & N. S. Dorfman (Eds.), *Economics of the environment: Selected readings* (3rd ed., pp. 5–19). New York: W. W. Norton.

Harding, S. (1998). *Is science multicultural? Postcolonialism, feminism, and epistemologies*. Bloomington: University of Indiana Press.

Harwood, J. (1993). *Styles of scientific thought: The German genetics community, 1900–1933*. Minneapolis: University of Minnesota Press.

Hauser, M. D. (1996). *The evolution of communication*. Cambridge, MA: MIT Press.

Hayles, N. K. (1999). *How we became posthuman: Virtual bodies in cybernetics, literature, and informatics*. Chicago: University of Chicago Press.

Headrick, D. R. (1990). *The invisible weapon: Telecommunications and international relations, 1851–1945*. New York: Oxford University Press.

Headrick, D. R. (2000). *When information came of age: Technologies of knowledge in the age of reason and revolution, 1700–1850*. New York: Oxford University Press.

Health and Safety Executive. (1999). *Year 2000 risk assessment*. London: Author.

Healy, B. (1993). Statement before the House Small Business Subcommittee on Regulation, Business Opportunities and Energy, 103rd Congress, 1st Session. Testimony of Bernadine Healy, M.D., Director of the National Institutes of Health.

Herman, E. S., & Chomsky, N. (1988). *Manufacturing consent: The political economy of the mass media*. New York: Pantheon.

Herman, E. S., & McChesney, R. W. (1997). *The global media: The new missionaries of corporate capitalism*. Washington, DC: Cassell.

Herrera, S. (2000, March). Reversal of fortune. *Red Herring: The Business of Technology*, pp. 154–164.

Hill, L. D. (1990). *Grain grades and standards: Historical issues shaping the future*. Urbana: University of Illinois Press.

Hilts, P. J. (2000, November 1). Company tried to bar report that HIV vaccine failed. *New York Times*, p. A20.

Hindle, J. (1999, February). Science by press release. *The Times Higher Education Supplement*, pp. 10–11.

History of Bioinformatics. (2001). Cambridge, MA: History of Recent Science and Technology Program, Massachusetts Institute of Technology (http://hrst.mit.edu/hrs/, consulted July 15, 2001).

Hoban, T., Woodrum, E., & Czaja, R. (1992). Public opposition to genetic engineering. *Rural Sociology, 57*, 476–493.

Hoffmeyer, J. (1997). Biosemiotics: Towards a new synthesis in biology. *European Journal for Semiotic Studies, 9*(2), 355–376.

Hookway, B. (1999). *Pandemonium: The rise of predatory locales in the postwar world.* Princeton: Princeton Architectural Press.

Hornig, S. (1991). *Monsanto Corporation and bovine somatotropin: A case study in failed public relations.* College Station, TX: Texas A&M University Center for Biotechnology Policy and Ethics discussion paper CBPE 91-10.

House of Lords. (2000, February 23). *Science and society.* Select committee on Science and Technology, 3rd report.

Hubbard, R., & Wald, E. (1993). *Exploding the gene myth: How genetic information is produced and manipulated by scientists, physicians, employers, insurance companies, educators, and law enforcers.* Boston: Beacon.

Humphreys, D., Eggan, K., Akutsu, H., Hochedlinger, K., Rideout, W. M. III, Biniszkiewicz, D., Yanagimachi, R., & Jaenisch, R. (2001). Epigenetic instability in ES cells and cloned mice. *Science, 293,* 95–97.

Humphrys, J. (2002). *The great food gamble.* London: Coronet Books.

Huxford, J. (2001). Beyond the referential: Uses of visual symbolism in the press. *Journalism, 2*(1), 45–71.

Huxley, A. (1989a [1932]). *Brave new world.* New York: Perennial Library.

Huxley, A. (1989b). *Brave new world revisited.* New York: Perennial Library.

The Independent. (1999, February 15). The merits of letting us know what we are eating, p. 3.

Irving, C. (1999, Fall). UC Berkeley's experiment in research funding. *National Crosstalk, 7*(4), 14–16.

Jewish Gothic. (2001). Frankenstein and the Golem, p. 1 (www.jewishgothic.com/golem.html).

Johnstone, R. A., & Grafen, A. (1993). Dishonesty and the handicap principle. *Animal Behavior, 46,* 759–764.

Johnstone, R. A., & Norris, K. (1993). Badges of status and the cost of aggression. *Behavioral Ecology and Sociobiology, 32,* 127–134.

Kahin, B. (2001). The expansion of the patent system: Politics and political economy. *First Monday, 6*(1) (http://firstmonday.org/).

Kahin, B. (2003). Codification in context. In S. Braman (Ed.), *The emergent global information policy regime.* Houndsmills, UK: Palgrave Macmillan.

Kasperson, R. E., et al. (1988). The social amplification of risk: A conceptual framework. *Risk Analysis, 8*(2), 177–204.

Kass, L., & Wilson, J. Q. (1998). *The ethics of human cloning.* Washington, DC: AEI Press.

Keller, E. F. (1983). *A feeling for the organism: The life and work of Barbara McClintock.* San Francisco: W. H. Freeman.

Kelso, P. (2000, September 21). Greenpeace wins key GM case. *The Guardian,* p. 1.

Kennedy, D. (1982). The social sponsorship of innovation. *Technology in Society, 4,* 253–265.

Kenney, M. (1986). *Biotechnology: The university–industrial complex.* New Haven and London: Yale University Press.

Kerbo, H. R. (1991). *Social stratification and inequality.* New York: McGraw-Hill.

Kevles, D. J., & Hood, L. (1992). Reflections. In D. J. Kevles & L. Hood (Eds.), *The code of codes: Scientific and social issues in the human genome project* (pp. 300–328). Cambridge, MA: Harvard University Press.

Kiernan, V. (1999, January 8). Creating a genetic Rosetta stone. *The Chronicle of Higher Education,* pp. A19–A20.

King, R. T. (1996, April 25). Bitter pill: How a drug firm paid for university study, then undermined it. *Wall Street Journal,* p. A1.

King, S. R. (1994). Establishing reciprocity: Biodiversity, conservation and new models for cooperation between forest-dwelling peoples and the pharmaceutical industry. In T. Greaves (Ed.), *Intellectual property rights for indigenous peoples: A sourcebook* (pp. 69–82). Oklahoma City, OK: Society for Applied Anthropology.

Kittay, E. F. (1987). *Metaphor: Its cognitive force and linguistic structure.* Oxford, UK: Oxford University Press.

Kitzinger, J., & Reilly, J. (1997). The rise and fall of risk reporting: Media coverage of human genetics research, "false memory syndrome" and "mad cow disease." *Electronic Journal of Communication, 12*(3), 319–350.

Klein, H. (2003). Private governance for global communications: Technology, contracts and the internet. In S. Braman (Ed.), *The emergent global information policy regime.* Houndsmills, UK: Palgrave Macmillan.

Kloppenburg, J., Jr. (1988). *First the seed: The political economy of plant biotechnology, 1492–2000.* Cambridge: Cambridge University Press.

Kloppenburg, J., Jr., & Burrows, B. (1996). Biotechnology to the rescue? Twelve reasons why biotechnology is incompatible with sustainable agriculture. *The Ecologist, 26*(2), 61–68.

Kloppenburg, J., Jr., & Kleinman, D. L. (1988). Plant genetic resources: The common bowl. In J. R. Kloppenburg, Jr. (Ed.), *Seeds and sovereignty: The use and control of plant genetic resources* (pp. 1–18). Durham, NC: Duke University Press.

Knorr-Cetina, M. K. (1997). Sociality with objects: Social relations in postsocial knowledge societies. *Theory, Culture & Society, 14*(4), 1–30.

Konopka, J., & Hanson, J. (Eds.). (1985). *Information handling systems for genebank management.* Rome: IBGPR Secretariat.

Kreps, D. (1990). Corporate culture and economic theory. In J. Alt & K. Shepsle (Eds.), *Perspectives on positive political economy* (pp. 90–143). New York: Cambridge University Press.

Krimsky, S. (1982). *Genetic alchemy: The social history of the recombinant DNA controversy.* Cambridge, MA: MIT Press.

Krimsky, S. (1988). University entrepreneurship and the public purpose. In P. DeForest, M. S. Frankel, J. S. Poindexter, & V. Weil (Eds.), *Biotechnology: Professional issues and social concerns* (pp. 34–42). Washington, DC: American Association for the Advancement of Science.

Krimsky, S. (1991). *Biotechnics & society: The rise of industrial genetics.* New York: Praeger.

Krimsky, S., Ennis, J. G., & Weissman, R. (1991). Academic–corporate ties in biotechnology: A quantitative study. *Science, Technology & Human Values, 16*(3), 275–287.

Kroker, A. (1992). *The possessed individual: Technology and the French postmodern.* New York: St. Martin's Press.

Kull, S., & Ramsay, C. (2000). Elite misperceptions of U.S. public opinion and foreign policy. In B. L. Nacos, R. Y. Shapiro, & P. Isernia (Eds.), *Decisionmaking in a glass house* (pp. 95–110). Lanham, MD: Rowman & Littlefield.

Laird, S. (1994). Natural products and the commercialization of traditional knowledge. In T. Greaves (Ed.), *Intellectual property rights for indigenous peoples: A sourcebook* (pp. 145–162). Oklahoma City, OK: Society for Applied Anthropology.

Lakoff, G., & Johnson, M. (1980). *Metaphors we live by.* Chicago: University of Chicago Press.

Lakoff, G., & Nunez, R. E. (2000). *Where mathematics comes from.* New York: Basic Books.

Lakoff, G., & Turner, M. (1989). *More than cool reason: A field guide to poetic metaphor.* Chicago: University of Chicago Press.

Latour, B. (1988). *The pasteurization of France* (J. Law & A. Sheridan, Trans.). Cambridge, MA: Harvard University Press.

Lazonick, W. (1991). *Business organization and the myth of the market economy.* Cambridge: Cambridge University Press.

Le Bon, G. (1895/1969). *The crowd: A study of the popular mind.* New York: Ballantine.

Leahy, P. J., & Mazur, A. (1980). The rise and fall of public opposition in specific social movements. *Social Studies of Science, 10*(3), 259–284.

Lean, G. (1999, March 7). How I told the truth and was sacked. *The Independent on Sunday,* p. 7.

Lean, G. (2003, February 26). GM crops are breeding with plants in the wild. *The Independent,* p. 1 (independent.co.uk).

Leese, M. (1996). Is an American mouse a European mouse? Towards a sociology of patents. In A. Webster & K. Parker (Eds.), *Innovation and the intellectual property system* (pp. 171–191). London: Kluwer Law International.

Leeuwis, C. (1993). *Of computers myths and modelling*. Wageningen, The Netherlands: Landbouwuniversiteit.

Leith, J. A. (1978). The war of images surrounding the Commune. In J. A. Leith (Ed.), *Images of the commune* (pp. 101–150). Montreal: McGill-Queen's University Press.

Lessig, L. (1999). *Code and other laws of cyberspace*. New York: Basic Books.

Levine, I., & Brown, L. R. (1970). *Knock three times*. Pocket Full of Tunes.

Lewis, L. (2000, October 29). Job testing may create a genetic underclass. *The Independent on Sunday*, Business section, p. 5.

Liebes, T., & Katz, E. (1999). *Export of meaning: Cross-cultural readings of Dallas*. London: Blackwell.

Lievrouw, L. A. (1986). *The communication network as interpretive environment: "Sense-making" among biomedical research scientists*. Unpublished dissertation, University of Southern California, Los Angeles, CA.

Lievrouw, L. A. (2002). Determination and contingency in new media development: Diffusion of innovations and social shaping of technology perspectives. In L. A. Lievrouw & S. Livingstone (Eds.), *Handbook of new media: Social shaping and consequences of ICTs* (pp. 183–199). London: Sage.

Litman, J. (1994). Mickey Mouse emeritus: Character protection and the public domain. *University of Miami Entertainment and Sports Law Review, 11*, 429.

Locke, J. (1690/1998). *An essay concerning human understanding*. New York: Viking Penguin.

Longworth, R. C. (1992, Oct. 27). "Soybean war" mushrooming into global trade crisis. *Chicago Tribune*, Sec. 3, p. 1.

Lueck, D. L. (1995). The economic organization of wildlife institutions. In T. L. Anderson & P. J. Hill (Eds.), *Wildlife in the marketplace* (pp. 1–24). Lanham, MD: Rowman & Littlefield.

Lyotard, J. F. (1984). *The postmodern condition: A report on knowledge*. Minneapolis: University of Minnesota Press.

MacCordy, E. L. (1988). The impact of proprietary arrangements on universities. In P. DeForest, M. S. Frankel, J. S. Poindexter, & V. Weil (Eds.), *Biotechnology: Professional issues and social concerns* (pp. 12–19). Washington, DC: American Association for the Advancement of Science.

Macksey, K. (1989). *For want of a nail: The impact of war on logistics and communications*. London: Brassey's (UK) Ltd.

Magat, W. A., & Viscusi, W. K. (1992). *Informational approaches to regulation*. Cambridge: MIT Press.

Mahoney, P. G. (2001, December). Norms and signals: Some skeptical observations. *University of Richmond Law Review, 36*, 387–406.

Mamiya, C. J. (1992). *Pop art and consumer culture: American super market*. Austin: University of Texas Press.

Marcuse, H. (1964). *One-dimensional man: Studies in the ideology of advanced industrial society*. Boston: Beacon.

Marcuse, H. (1969). *An essay on liberation*. Boston: Beacon.

Martin, B. (1999). Suppressing research data: Methods, context, accountability and responses. *Accountability in Research, 6*, 333–372 (http://www.uow.edu.au/arts/sts/, consulted July 11, 2001).

Martinson, O. B., & Campbell, G. R. (1980). Betwixt and between: Farmers and the marketing of agricultural inputs and outputs. In F. H. Buttel & H. Newby (Eds.), *The rural sociology of the advanced societies: Critical perspectives* (pp. 215–254). London: Croom Helm.

Maskus, K. E. (2000). *Intellectual property rights in the global economy*. Washington, DC: Institute for International Economics.

May, C. (1998). Thinking, buying, selling: Intellectual property rights in political economy. *New Political Economy, 3*(1), 59–78.

May, C. (2000). *A global political economy of intellectual property rights: The new enclosures?* London: Routledge.

Mayer, L. V. (1986). Farm exports and the farm economy: Economic and political interdependence. In W. P. Browne & D. F. Hadwiger (Eds.), *World food policies: Toward agricultural independence* (pp. 17–27). Boulder: Lynne Riemer.

Maynard-Smith, J. (1991). Honest signalling; the Philip Sydney game. *Animal Behavior, 42,* 1034–1035.

Maynard-Smith, J. (1994). Must reliable signals always be costly? *Animal Behavior, 47,* 1115–1120.

McCorkle, C. M. (Ed.). (1989). *The social sciences in international agricultural research: Lessons from the CRSPs.* London: Lynne Riemer.

McCrary, S. V., Anderson, C. B., Jakovljevic, J., Khan, T., McCullough, L. B., Wray, N. P., & Brody, B. A. (2000). A national survey of policies on disclosure of conflicts of interest in biomedical research. *New England Journal of Medicine, 343*(22), 1621–1626.

McGuigan, J. (2000). British identity and "the people's princess." *The Sociological Review, 48*(1), 1–18.

McHughen, A. (2000). *A consumer's guide to GM food: From green genes to red herrings.* Oxford: Oxford University Press.

McKibben, B. (1996, July 11). Some versions of pastoral. *The New York Review of Books,* pp. 42–45.

McKie, R. (1999, October 17). Why Britain's scientific establishment got so ratty with this gentle boffin. *The Observer,* p. 10.

McLuhan, M. (1964). *Understanding media: The extensions of man.* New York: Macmillan.

McLuhan, M. (1968). *Through the vanishing point.* New York: HarperCollins.

McLuhan, M., & Fiore, Q. (1968). *War and peace in the global village.* New York: McGraw-Hill.

McNally, R., & Wheale, P. (1996). Biopatenting and biodiversity: Comparative advantages in the new global order. *The Ecologist, 26*(5), 222–229.

Meek, J. (2000, May 20). Prince's luxury lifestyle conflicts with image of ascetic and reluctant royal. *The Guardian,* p. 3.

Meikel, J. (1999, June 18). "Buffer zones" no GM safeguard. *The Guardian,* p. 6.

Melchett, P. (1999, August 1). Today's vandal will be tomorrow's hero. *The Independent,* p. 24.

Merges, R. P. (1996). Property rights theory and the commons: The case of scientific research. *Social Philosophy and Policy, 13*(2), 145–167.

Merton, R. (1948). The self-fulfilling prophecy. *The Antioch Review, 8,* 193–218.

Michael, M. (1996). Ignoring science: Discourses of ignorance in the public understanding of science. In A. Irwin & B. Wynne (Eds.), *Misunderstanding science?* (pp. 107–125). New York: Cambridge University Press.

Miller, D. (1976). *Social justice.* Oxford: Clarendon.

Miller, H. I., & Huttner, S. L. (1995). Food produced with new biotechnology: Can labeling be anti-consumer? *Journal of Public Policy & Marketing, 14*(2), 330–333.

Misakian, A. L., & Bero, L. A. (1998). Publication bias and research on passive smoking: Comparison of published and unpublished studies. *Journal of the American Medical Association, 280*(3), 250–253.

Mitroff, I. I. (1974). *The subjective side of Science: A philosophical inquiry into the psychology of the Apollo moon scientists.* Amsterdam: Elsevier.

Monbiot, G. (2001a). *Captive state: The corporate takeover of Britain.* London: Pan Books.

Monbiot, G. (2001b, November 19). The misappliance of science. *The Observer,* p. 30.

Moody, G. (2002, January 10). Free software survives downturn. *The Guardian On Line,* pp. 1–3.

Mooney, P. R. (1988). Biotechnology and the North–South conflict. In *Biotechnology revolution and the third world: Challenges and policy options* (pp. 253–278). New Delhi: Research and Information Systems.

Moore, P. J., Reagan-Wallin, N. L., Haynes, K. F., & Moore, A. J. (1997). Odour conveys status on cockroaches. *Nature, 389,* 25.

Morgan, O. (2000, April 16). Biotech's gene blueprint. *The Observer*, Business section, p. 4.

Mulkay, M. (1976). Norms and ideology in science. *Social Science Information, 15*, 637–656.

Mulkay, M. (1997). *The embryo research debate: Science and the politics of reproduction.* Cambridge: University Press.

Mumford, L. (1934). *Technics and civilization.* New York: Harcourt, Brace.

Murdock, G. (1997). The re-enchantment of the world: Religion and the transformations of modernity. In S. M. Hoover & K. Lundby (Eds.), *Rethinking media, religion, and culture* (pp. 85–101). London: Sage.

Murdock, G., Petts, J., & Horlick-Jones, T. (2003). After amplification: Rethinking risk communication. In N. Pigeon, R. Kasperson, & P. Slovic (Eds.), *The social amplification of risk.* Cambridge: Cambridge University Press.

Myerson, G. (2000). *Donna Haraway and GM foods.* Cambridge: Icon.

Myerson, G. (2001). *Ecology and the end of postmodernity.* Cambridge: Icon.

Naficy, H. (1996). Phobic spaces and liminal panics: Independent transnational film genre. In R. Wilson & W. Dissanayake (Eds.), *Global/local: Cultural production and the transnational imaginary* (pp. 119–144). Durham, NC: Duke University Press.

Narayanan, N. (2002, May 8). AAAS website seeks to protect traditional knowledge of plants (www.aaas.org/news/releases/2002/0508tek.shtml).

NASA breakthrough method may lead to smaller electronics. (2002, November 25). NASA press release.

National Science Board. (2000). *Science and engineering indicators—2000.* Arlington, VA: National Science Foundation (NS8-00-1).

"Needed: Seed money." (1994, September 10). *The Economist, 332*(7880), 100.

Nelkin, D. (1982, May 14). Intellectual property: The control of scientific information. *Science, 216*, 704–708.

Nelkin, D. (1995). *Selling science: How the press covers science and technology* (2nd ed.). New York: W. H. Freeman.

Nelkin, D., & Lindee, M. S. (1995). *The DNA mystique: The gene as a cultural icon.* New York: W. H. Freeman.

Nelson, P. (1974). Advertising as information. *Journal of Political Economy, 81*, 729–754.

Nelson, P. (1975). The economic consequences of advertising. *Journal of Business, 48*(2), 213–341.

Neuman, W. R. (1991). *The future of the mass audience.* New York: Cambridge University Press.

New York Times. (2000, June 27). Genetic code of human life is cracked by scientists, p. A1.

Nisbet, M., & Lewenstein, B. V. (2000). *Content analysis of biotechnology, 1970–1994.* Unpublished manuscript, Cornell University, Department of Communication, Ithaca, NY.

Nisbet, M., & Lewenstein, B. V. (2001, May). *A comparison of U.S. media coverage of biotechnology with public perceptions of genetic engineering, 1995 to 1999.* Paper presented to International Communication Association, Washington, DC.

Noelle-Neumann, E. (1984). *The spiral of silence: Public opinion, our social skin.* Chicago: University of Chicago Press.

Nora, S., & Minc, A. (1980). *The computerization of society.* Cambridge: MIT Press.

North, D. C. (1990). *Institutions, institutional change and economic performance.* Cambridge: Cambridge University Press.

"Not a grain of truth: Cambodia." (1992, January 11). *The Economist, 322*(7441), 34.

Novak, M. (1997). Transmitting architecture: The transphysical city. In A. Kroker & M. Kroker (Eds.), *Digital delirium* (pp. 260–271). New York: St. Martin's Press.

Office of Technology Assessment. (1982). *Genetic technology: A new frontier.* Boulder, CO: Westview.

Office of Technology Assessment. (1984, January). *Commercial biotechnology: An international analysis.* OTA Report, Washington, DC: U.S. Government Printing Office, OTA-BA-218 (http://www.wws.princeton.edu/~ota/ns20/pubs_f.html, consulted July 15, 2001).

Office of Technology Assessment. (1988). *New developments in biotechnology: U.S. investment in biotechnology.* Washington, DC: U.S. Government Printing Office, OTA-BA-360.

Old MacDonald had an option: Crop insurance. (1995, February 11). *The Economist, 334*(7901), p. 68.

Oliveira, O. S. (1992). Food, advertising, and hunger. *Directions, 6*(1), 9–12.

Organization for Economic Cooperation and Development. (1988). *Biotechnology and the changing role of government.* Paris: Author.

Organization for Economic Cooperation and Development. (1989). *Biotechnology: Economic and wider impacts.* Paris: Author.

Oyama, S. (2000). *The ontogeny of information: Developmental systems and evolution* (2nd ed.). Durham, NC: Duke University Press.

Panic in the petri dish. (1994, July 23). *The Economist, 332*(7873), 61–62.

Partnership perils. (2001, June 15). *Science, 292*(5524), 1983.

Pearton, M. (1984). *Diplomacy, war and technology since 1830.* Lawrence: University Press of Kansas.

Peering into 2010: A survey on the future of medicine. (1994, March 19). *The Economist, 330*(7855), p. 13.

Pence, G. (1998). *Who's afraid of human cloning?* Lanham, MD: Rowman & Littlefield.

Peters, J. D. (1988). Information: Notes toward a critical history. *Journal of Communication Inquiry, 12*(2), 9–23.

Petit, C. W. (1998). Germinating access. *U.S. News and World Report, 125*, 60.

Phillips, M. J. (1989). Biotechnology: A new frontier for agriculture. In M. Drabenstott (Ed.), *Positioning agriculture for the 1990s: A new decade of change* (pp. 45–62). Washington, DC: National Planning Association.

Plein, L. C. (1991). Popularizing biotechnology: The influence of issue definition. *Science, Technology, & Human Values, 16*(4), 474–490.

Plomin, J. (2001, November 6). Keeping the faith. *Guardian Education*, p. 31.

Polanyi, K. (1944/1957). *The great transformation: The political and economic origins of our time.* Boston: Beacon.

Pool, I. de Sola. (1983). *Technologies of freedom.* Cambridge: Belknap.

Posner, E. A. (2000). *Law and social norms.* Cambridge, MA: Harvard University Press.

Poulter, S. (1999, February 27). Top-brand foods made with mutant crops, and the companies don't even know it. *Daily Mail*, p. 7.

Powell, W. W. (1996). Inter-organizational collaboration in the biotechnology industry. *Journal of Institutional and Theoretical Economics, 152*(1), 197–215.

Price, D. J. de Solla. (1963). *Little science, big science.* New York: Columbia University Press.

Price, D. J. de Solla. (1988). Bridging the gap between academia and industry: The scientist's role. In P. DeForest, M. S. Frankel, J. S. Poindexter, & V. Weil (Eds.), *Biotechnology: Professional issues and social concerns* (pp. 20–27). Washington, DC: American Association for the Advancement of Science.

Priest, S. H. (1995). Information equity, public understanding of science, and the biotechnology debate. *Journal of Communication, 45*(1), 39–54.

Priest, S. H. (2001a). *A grain of truth.* New York: Rowman & Littlefield.

Priest, S. H. (2001b). Cloning: A study in news production. *Public Understanding of Science, 10*(1), 59–69.

Priest, S. H. (2001c). Misplaced faith: Communication variables as predictors of encouragement for biotechnology development. *Science Communication, 23*(2), 97–110.

Priest, S. (2002, September). U.S. public opinion divided over biotechnology? *Nature Biotechnology, 13*, 939–942.

Priest, S., Bonfadelli, H., & Rusanen, M. (in press). The trust gap hypothesis: Predicting support for biotechnology across national cultures: As a function of trust in actors. *Risk Analysis.*

Priest, S. H., & Talbert, J. (1994). Mass media and the ultimate technological fix: Newspaper coverage of biotechnology. *Southwestern Mass Communication Journal, 10*(1), 76–85.

Primrose, S. B. (1991). *Molecular biotechnology* (2nd ed.). Oxford and London: Blackwell Scientific.

The Prince of Wales. (1999, June 1). Genetically modified food: The Prince of Wales asks: Is it an innovation we can do without? (www.princeofwales.gov.uk/forum).

"Privacy versus database statistics." (1991, November 16). *Science News, 140*(20), 315.

Purdue, D. A. (2000). *Anti-GenetiX: The emergence of the anti-GM movement.* Aldershot: Ashgate.

Quijano, A., & Wallerstein, I. (1992). Americanity as a concept, or the Americas in the modern world-system. *International Social Science Journal, 13*(4), 549–557.

Rabin, R. (1987). Sustaining American leadership in biotechnology. In I. K. Vasil (Ed.), *Biotechnology: Perspectives, policies, and issues* (pp. 229–234). Gainesville: University of Florida Press.

Rabinow, P. (1996). *Making PCR: A story of biotechnology.* Chicago, IL: University of Chicago Press.

Radford, T. (1999a, February 23). They don't know, you know. *The Guardian,* p. 17.

Radford, T. (1999b, March 9). "No regret" at alarm on GM food. *The Guardian,* p. 15.

Rasmussen, E. B., & Perri, T. J. (2001). Can high prices ensure product quality when buyers do not know the sellers' cost? *Economic Inquiry, 39*(4), 561–567.

Reiss, M. J., & Straughan, R. (1996). *Improving nature? The science and ethics of genetic engineering.* New York: Cambridge University Press.

Renn, O. (1991). Risk communication and the social amplification of risk. In R. E. Kasperson & P. J. M. Stallen (Eds.), *Communicating risks to the public: International perspectives* (pp. 287–324). Dordrecht: Kluwer Academic.

Richon, A. B. (2001). *A short history of bioinformatics* (http://www.netsci.org/Science/Bioinform/, consulted July 15, 2001).

Ridley, M. (2001). *The cooperative gene: How Mendel's demon explains the evolution of complex beings.* New York: The Free Press.

Rifkin, J. (1984). *Algeny: A new world.* New York: Viking Penguin.

Rifkin, J. (1998). *The biotech century: Harnessing the gene and remaking the world.* New York: Tarcher/Putnam.

Rifkin, J. (1999). *The biotech century: How genetic commerce will change the world.* London: Phoenix.

Riley, J. G. (2001, June). Silver signals: Twenty-five years of screening and signaling. *Journal of Economic Literature, 39,* 432–478.

Rissler, J., & Mellon, M. (1996). *The ecological risks of engineered crops.* Cambridge, MA: The MIT Press.

Ritchie, L. D. (1986). Shannon and Weaver: Unraveling the paradox of information. *Communication Research, 13*(2), 278–298.

Ritchie, L. D. (1991). *Information.* Newbury Park, CA: Sage.

Ritchie, L. D. (2001). *"Fat and happy," "passed away," and "birds of prey": Three arguments for a more complex interpretation of metaphors.* Unpublished paper, Portland State University, Portland, OR.

Ritchie, L. D. (2002). Monastery or economic enterprise: Opposing or complementary metaphors of higher education? *Metaphor and Symbol, 17*(1), 45–55.

Ritchie, L. D. (2003). Statistical probability as a metaphor for epistemological probability. *Metaphor and Symbol, 18,* 1–11.

Roberts, M. K., Schwartz, C. M., Stohl, M. S., & Targ, H. R. (1986). The policy consequences of the Green Revolution: The Latin American case. In W. P. Browne & D. F. Hadwiger (Eds.), *World food policies: Toward agricultural independence* (pp. 137–151). Boulder: Lynne Riener.

Roobeek, A. J. M. (1990). *Beyond the technology race: An analysis of technology policy in seven industrial countries.* Amsterdam/New York: Elsevier Science.

Rosenberg, J. A. (1996, February 8). Secrecy in medical research. *New England Journal of Medicine, 334,* 392–394.

Rosenzweig, R. M. (1999, Spring). A good deal for all? *National Crosstalk, 7*(2), 11–2.

The Royal Society. (2002). *Genetically modified plants for food use and human health—an update* (Policy Document 4/02). London: Author.

Rudolph, F. B., & McIntire, L. V. (Eds.). (1996). *Biotechnology: Science, engineering and ethical challenges for the 21st century.* Washington, DC: Joseph Henry Press.

Ruppert, D. (1994). Buying secrets: Federal government procurement of intellectual cultural property. In T. Greaves (Ed.), *Intellectual property rights for indigenous peoples: A sourcebook* (pp. 111–128). Oklahoma City, OK: Society for Applied Anthropology.

Russell, S. A. (2002, April). Talking plants. *Discover,* pp. 47–50.

Sainbury, D. (1999, May 17). Don't knock the boffins. *New Statesman,* p. 28.

Sapp, S. G., & Harrod, W. J. (1990). Consumer acceptance of irradiated food: A study of symbolic adoption. *Journal of Consumer Studies and Home Economics, 14*(2), 133–145.

Saracevic, T., & Kesselman, M. (1993). Trends in biotechnology information and networks: Implications for policy. In G. T. Tzotzos (Ed.), *Biotechnology R&D trends: Science policy for development, Annals of the New York Academy of Sciences, 700* (pp. 135–144). New York: New York Academy of Sciences.

Scazzieri, R. (1993). *A theory of production: Tasks, processes, and technical practices.* Oxford: Clarendon.

Schement, J. R., & Lievrouw, L. A. (1987). *Competing visions, complex realities: Social aspects of the information society.* Norwood, NJ: Ablex.

Schramm, W. (1955). Information theory and mass communication. *Journalism Quarterly, 32,* 131–146.

Schuh, G. E. (1986). Maximizing US benefits from agricultural independence. In W. P. Browne & D. F. Hadwiger (Eds.), *World food policies: Toward agricultural independence* (pp. 29–41). Boulder, CO: Lynne Riener.

Schuh, G. E. (1989). Global factors affecting US markets in the next decade. In Drabenstott, *op cit.,* pp. 31–41.

Scott, T. (Director). (1995). *Crimson tide.* A Panavision film.

Sedjo, R. A. (1992). Property rights, genetic resources and biotechnological change. *Journal of Law and Economics, 35*(1), 199–213.

Sedjo, R. A., & Simpson, R. D. (1995). Property rights contracting and the commercialization of biodiversity. In T. L. Anderson & P. J. Hill (Eds.), *Wildlife in the marketplace* (pp. 167–178). Lanham, MD: Rowman & Littlefield.

Sell, S. K. (1998). *Power and ideas: North–South politics of intellectual property and antitrust.* Albany, NY: State University of New York Press.

Sell, S. K., & May, C. (2001). Moments in law: Contestation and settlement in the history of intellectual property. *Review of International Political Economy, 8*(3), 467–500.

Shannon, C. (1949). The mathematical theory of communication. In C. Shannon & W. Weaver (Eds.), *The mathematical theory of communication* (pp. 3–91). Urbana, IL: University of Illinois Press.

Shannon, C. (1956). The bandwagon. *IRE Transactions on Information Theory, 2,* 3.

Shiva, V. (1998). *Biopiracy: The plunder of nature and knowledge.* Dartington: Green Books.

Shiva, V. (2000). *Stolen harvest: The hijacking of the global food supply.* Cambridge, MA: South End Press.

Siddhanti, S. K. (1991). *Multiple perspectives on risk and regulation: The case of deliberate release of genetically engineered organisms into the environment.* New York/London: Garland.

Silver, B. L. (1998). *The ascent of science.* New York: Oxford University Press.

Simring, F. R. (1978, June 28). Letter to the Editor. *The New York Times,* p. 22.

Singer, P. (1975). *Animal liberation.* New York: Avon.

Sklair, L. (1998). As political actors. *New Political Economy, 3*(2), 284–288.

Slovic, P. (1987). *Preference reversals.* College Park, MD: Center for Philosophy and Public Policy, University of Maryland.

Smith, J. E. (1985). *Biotechnology principles.* Washington, DC: American Society for Microbiology.

Spence, A. M. (1973). Job market signaling. *Quarterly Journal of Economics, 87*(3), 355–374.

Spence, A. M. (1974). *Market signaling: Informational transfer in hiring and related screening processes*. Cambridge, MA: Harvard University Press.

Spitz, P. (1983). *Food systems and society in India*. Geneva: UN Research Institute for Social Development.

Srinivas, K. R. (2002). The case for biolinuxes and other pro-commons initiatives. *Sarai Reader 2002*, pp. 321–328.

Stape, A. (2001, July 12). Turf wars: Eco-terrorists threaten biotech companies. *Providence Journal-Bulletin*, p. E1.

Star, S. L., & Bowker, G. (2002). How to infrastructure. In L. A. Lievrouw & S. Livingstone (Eds.), *Handbook of new media: Social shaping and consequences of ICTs* (pp. 151–162). London: Sage.

Stelfox, H. T., Chua, G., O'Rourke, K., & Detsky, A. S. (1998, January 8). Conflict of interest in the debate over calcium-channel antagonists. *New England Journal of Medicine, 338*(2), 101–106.

Stephenson, D. J., Jr. (1994). A legal paradigm for protecting traditional knowledge. In T. Greaves (Ed.), *Intellectual property rights for indigenous peoples: A sourcebook* (pp. 179–189). Oklahoma City, OK: Society for Applied Anthropology.

Steyer, R. (1994, November 21). Confusion: Wisconsin consumers face array of choices. *St. Louis Post-Dispatch*, p. 15.

Stiglitz, J. E. (1999). Knowledge as a global public good. In I. Kaul, I. Grunberg, & M. A. Stern (Eds.), *Global public goods: International co-operation in the 21st century* (pp. 308–325). New York: United Nations Development Programme/Oxford University Press.

Stone, A. R. (1995). *War of desire and technology at the close of the mechanical age*. Cambridge: MIT Press.

Stone, J. L. (1991, May). *Contextualizing biogenetic engineering and reproductive technologies: A question of communication, culture and criticism*. Paper presented to the International Communications Association, Chicago.

Stonich, S. C., & Dewalt, B. R. (1989). *The political economy of agricultural growth and rural transformation in Honduras and Mexico*. Boulder: Westview.

Strobel, G. (1993, December 18 & 25). Seeds in need: The Vavilov Institute. *Science News, 144*(25–26), 416–417.

Sulston, J., & Ferry, G. (2002). *The common thread*. London: Bantam.

Sunderland, N. (2002, July). *A framework for using biotechnology as media*. Presented to the conference, Towards Humane Technologies: Biotechnology, New Media, and Citizenship, Ipswich, Queensland, Australia, The University of Queensland.

Suzuki, D., & Knudtson, P. (1989). *Genetics: The clash between the new genetics and human values*. Cambridge: Harvard University Press.

Tarrant, J. (1992). Agriculture and the state. In I. R. Bowler (Ed.), *The geography of agriculture in developed market economies* (pp. 239–274). New York: Longman Scientific and Technical.

Taverne, D. (2001). Organic food and anti-science. *Prospect, 67*, 54–58.

Teitelman, R. (1989). *Gene dreams: Wall Street, academia, and the rise of biotechnology*. New York: Basic Books.

Ten Eyck, T. A. (1999). Shaping a food safety debate: Control efforts of reporters and sources in the food irradiation controversy. *Science Communication, 20*, 426–447.

Ten Eyck, T. A., Thompson, P. B., & Priest, S. H. (2001). Biotechnology in the United States of America: Mad or moral science? In G. Gaskell & M. W. Bauer (Eds.), *Biotechnology, 1996–2000: The years of controversy* (pp. 307–318). London: Science Museum.

Tengelin, V. (1981). The vulnerability of the computerised society. In H. P. Gassmann (Ed.), *Information, communication, and computer policies for the '80s* (pp. 205–213). Amsterdam: North-Holland.

Thomas, S. M. (1997, August 21). Letter. *Nature*, pp. 805–808.

Thomas, S. M. (1999). Genomics and intellectual property rights. *Drug Discovery Today, 4*(3), 134–138.

Thompson, P. B. (1997). *Food biotechnology in ethical perspective*. New York: Blackie Academic.

Thorbecke, E. (1992). *The anatomy of agricultural product markets and transactions in developing countries*. Washington, DC: Institute for Policy Reform.

Timmins, J., & George, N. (1993, July 4). Feingold, Kohl wrong to oppose BST. *Wisconsin State Journal*, p. 6E.

Tirohl, B. (2000). The photo-journalist and the changing news image. *New Media and Society, 2*(3), 335–352.

Toolis, K. (2000, May 5). DNA: It's war. *The Guardian*, Weekend section, pp. 8–19.

Traill, B. (1988). *Technology and food: Aims and findings of the EC FAST Programme's research into the prospects and needs of the European food system*. Brussels: European Commission.

Tuchman, G. (1978). *Making news: A study in the construction of reality*. New York: The Free Press.

Turkle, S. (1995). *Life on the screen: Identity in the age of the Internet*. New York: Simon & Schuster.

Turney, J. (1998). *Frankenstein's footsteps: Science, genetics and popular culture*. New Haven, CT: Yale University Press.

TVA v Hill, 437 US 153 (1978).

Unseemly couplings: Biotechnology mergers. (1995, May 13). *The Economist, 335*(7914), 66–67.

Van Alstyne, M., & Brynjolfsson, E. (1996, November 29). Could the Internet Balkanize science? *Science, 274*(5292), 1479–1480.

Van Brundt, J. (2000, September 15). Biotech patent fights. *Signals* (http://www.signalsmag.com, consulted March 12, 2001).

van Creveld, M. (1991). *Technology and war: From 2000 BC to the present* (rev. ed.). New York: The Free Press.

Van Dijck, J. (1998). *Imagenation: Popular images of genetics*. New York: New York University Press.

Van Wijk, J., Cohen, J. I., & Koem, J. (1993). *Intellectual property rights for agricultural biotechnology: Options and implications for developing countries*. The Hague: International Service for National Agricultural Research.

Vaver, D. (1990). Intellectual property today: Of myths and paradoxes. *Canadian Bar Review, 69*(1), 98–128.

Vernon, K. (1990). Pus, sewage, beer and milk: Microbiology in Britain, 1870–1940. *History of Science, 28*, 289–325.

Vervaeke, J., & Kennedy, J. M. (1996). Metaphors in language and thought: Falsification and multiple meanings. *Metaphor and Symbolic Activity, 11*(4), 273–284.

Vidal, J. (1999a, March 19). We're gagging on GM: Monsanto must face up to meltdown. *The Guardian*, p. 20.

Vidal, J. (1999b, October 7). Monsanto—We forgot to listen. *The Guardian*, p. 11.

Vidal, J. (2001a, August 8). Damages charge of 20p for anti-Gm crop protesters. *The Guardian*, p. 9.

Vidal, J. (2001b, November 30). Mexico's GM corn shocks scientists. *The Guardian*, p. 18.

Vonnegut, K. (1963). *Cat's cradle*. New York: Delacorte Press.

Waddell, C. (1990). The role of pathos in the decision-making process: A study in the rhetoric of science policy. *Quarterly Journal of Speech, 76*(4), 381–400.

Waging skyological warfare. (1995, January 28). *The Economist, 334*(7899), 57–58.

Waldron, J. (1993). From authors to copiers: Individual rights and social values in intellectual property. *Chicago-Kent Law Review, 68*, 841–887.

Walzer, M. (1983). *Spheres of justice: A defense of pluralism and equality*. Oxford: Martin Robinson.

Warde, I. (2001, March). For sale: U.S. academic integrity. *Le Monde Diplomatique* (http://www.transnationale.org/anglais/sources/information/, consulted July 11, 2001).

Watson, J. S. (1968/1980). *The double helix* (rev. ed.). New York: Norton.

Watt, N. (1999, June 7). Blair softens stance on GM foods. *The Guardian*, p. 1.

Weaver, W. (1949). Recent contributions to the mathematical theory of communication. In C. Shannon & W. Weaver (Eds.), *The mathematical theory of communication* (pp. 94–117). Urbana, IL: University of Illinois Press.

Webster, F. (1995). *Theories of the information society*. London: Routledge.

Weinberg, R. A. (1993). Reflections on the current state of data and reagent exchange among biomedical researchers: Panel on scientific responsibility and the conduct of research, Committee on Science, Engineering, and Public Policy. In *Responsible science: Ensuring the integrity of the research process, II* (pp. 66–78). Washington, DC: National Academy Press.

Weiss, R. (1999, February 3). Seeds of discord: Monsanto's gene police raise alarm on farmers' rights, rural tradition. *The Washington Post*, p. A01.

Wiegele, T. C. (1991). *Biotechnology and international relations: The political dimensions.* Gainesville: University of Florida Press.

Wilkie, T., & Graham, E. (1998). Power without responsibility: Media portrayals of Dolly and science. *Cambridge Quarterly of Healthcare Ethics, 7,* 150–159.

Wilkinson, J. (1987). *Europe within the world food system: Biotechnologies and new strategic options.* Brussels: EC FAST.

William, J. (1999, June 10). Consumer power forces food industry to modify approach. *Financial Times,* p. 11.

Williams, W. C. (1954). Asphodel, that greeny flower. In *The desert music and other poems.* New York: Random House.

Williamson, O. E. (1983, September). Credible commitments: Using hostages to support exchange. *American Economic Review, 73,* 519–540.

Wilmut, I., Campbell, K., & Tudge, C. (2000). *The second creation: Dolly and the age of biological control.* New York: Farrar, Straus & Giroux.

Wilson, L. K., & Virilio, P. (1996). Cyberwar, God and television: Interview with Paul Virilio. In T. Druckrey (Ed.), *Electronic culture: Technology and visual representation* (pp. 321–330). New York: Aperture.

Winston, B. (1998). *Media technology and society: A history.* London: Routledge.

Woodward, K. (1994). From virtual cyborgs to biological time bombs: Technocriticism and the material body. In G. Bender & T. Druckrey (Eds.), *Culture on the brink: Ideologies of technology* (pp. 47–64). Seattle: Bay Press.

Wuethrich, B. (1993, September 4). All rights reserved: How the gene-patenting race is affecting science. *Science News, 144,* 154–157.

Yoxen, E., & Hyde, B. (1987). *The social impact of biotechnology.* Manchester: European Foundation for the Improvement of Living and Working Conditions.

Zahavi, A. (1975). Mate selection—A selection for handicap. *Journal of Theoretical Biology, 53,* 205–214.

Zahavi, A. (1993). The fallacy of conventional signalling. *Phil. Trans. R. Soc. Lond. B, 338,* 227–230.

Zahavi, A., & Zahavi, A. (1997). *The handicap principle: A missing piece of Darwin's puzzle.* New York: Oxford University Press.

Zavarzadeh, M. (1976). *Mythopoeic reality: The postwar American nonfiction novel.* Urbana, IL: University of Illinois Press.

Zencey, E. (1991). Some brief speculations on the popularity of entropy as metaphor. *Metaphor and Symbolic Activity, 6*(1), 47–56.

Zittrain, J. (2003). Internet points of control. In S. Braman (Ed.), *The emergent global information policy regime.* Houndsmills, UK: Palgrave Macmillan.

Zuckerman, H. (1988). Intellectual property and diverse rights of ownership in science. *Science, Technology & Human Values, 13*(1&2), 7–16.

Zweiger, G. (2001). *Transducing the genome: Information, anarchy and revolution in the biomedical sciences.* New York: McGraw-Hill.

Author Index

Subject Index